American Institute of Chemical Engineers

Topical Conference on Information Technology Applications in the Chemical Process Industries 2004

Held at the 2004 AIChE Spring National Meeting

April 25-29, 2004
New Orleans, Louisiana, USA

Printed from e-media with permission by:

Curran Associates, Inc.
57 Morehouse Lane
Red Hook, NY 12571
www.proceedings.com

ISBN: 978-1-60423-474-9

Some format issues inherent in the e-media version may also appear in this print version.

American Institute of Chemical Engineers

Topical Conference on Information Technology Applications in the Chemical Process Industries
2004

TABLE OF CONTENTS

How to Use Information Technology to Optimize Benefits from Process Safety and Operability Information (PSOI)

V. Anthony Ciliberti, PE
Reliability Dynamics
vciliberti@reliabilitydynamics.com

Introduction

The petrochemical processing industries (CPI) today have a wealth of process safety and operability information (PSOI) at their disposal. Much of the information was developed or updated in the 1990s in compliance with governmental mandates, such as the OSHA standard 29 CFR 1910.119, Process Safety Management of Highly Hazardous Chemicals. Unfortunately, these industries are not fully utilizing their PSOI due to inefficiencies in data management. The result is increased risk and exposure.

PSOI is typically kept in a variety of media and data formats and is difficult to access, correlate, and maintain. Many electronic PSOI data sets do not contain discrete parameters for correlating them with other information. HAZOP studies, for example, use a nodal analysis method that fails to explicitly identify process equipment items and safeguards and their roles in fault, event, and consequence sequences. These and other related problems make it difficult for personnel to efficiently use PSOI in assessing and reducing risks associated with the different parts of their facilities.

In this paper, I propose the use of a hierarchal functional location structure to represent an enterprise in its entirety and identify risk levels at any point within that enterprise. I suggest methods for efficiently converting legacy PSOI into relational data sets and for dynamically interfacing multiple relational PSOI data sets with each other and enterprise data to concisely identify and present current risk levels, safeguards, and other applicable information. I demonstrate the advantage of using a universal risk matrix as one standard of measure for different risk assessment methodologies and datasets. I describe how to measure and determine the criticality of safeguards and use safeguard criticality to prioritize operations and maintenance activities.

A Need for Information

A stakeholder should be able to query PSOI data in such a manner that he or she could quickly answer such questions as:

- Where is there exposure in my facilities, and what are the current risk levels?
- What are the hazards? Where is there potential for production or economic loss? What are the scenarios for those hazards and/or economic losses?
- What hazardous chemicals are present and in what quantities?
- What incidents have occurred or have a potential for occurring?
- What procedures, inspections, and other safeguards are in place to mitigate risk? Are they effective in mitigating risk?

Most of the information required to answer these questions exists today in the processing industries in one form or another but is generally fragmented, disorganized, out-of-date, inaccessible, and/or non-relational. Rarely is all applicable PSOI pulled together for decision-making purposes because doing so would require a herculean manual effort. The good news is that today's information technology can make this information readily available.

Defining Common Reference Points

An enterprise must have a common system of reference for uniquely identifying all of its subsets. SAP AG's enterprise application software has defined such a field, "functional location," which is part of a hierarchal functional location structure that is client-configured during the implantation and setup of the SAP R/3 PM module. However, SAP software is not required to build a functional location structure.

To make it usable as described in this paper, the functional location structure must be configured so it captures all facets and components of a corporate business unit in a hierarchal manner, including administrative areas, processing areas, utilities, etc. Top-level functional locations identify facility areas, or functional area locations (FALs). Bottom-level functional locations identify individual equipment functions, or functional equipment locations (FELs). Each functional location is unique.

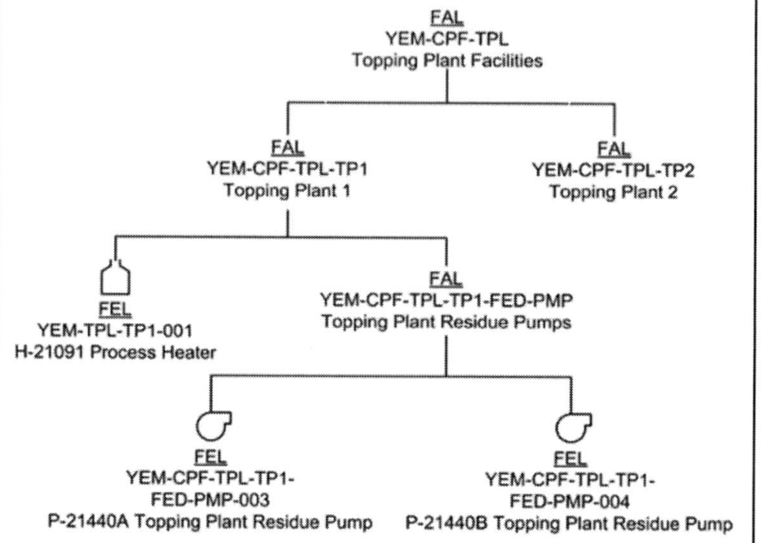

Figure 1 - Example of a Functional Location Structure.

It is important to distinguish between FELs and equipment. Functional location identifies the process function (e.g. FEL: YEM-CPF-TPL-TP1-FED-PMP-003, Topping Plant Residue "A" Pump, Tag No.: P-21440A, Spec: 168 GPM @ 160 PSI DP), whereas equipment identifies the physical asset performing that function (e.g. Equipment No. 3021976, Union Pumps Model 2X3-13C VLK, Serial# CH92290AX-1). Equipment items are installed in functional locations. The data relationship between functional location and equipment is one-to-many, but in most cases is a one-to-one ratio.

The hierarchal structure allows a stakeholder to review attribute data for any bottom-level FEL or to sum attribute data at higher-level "parent" functional locations. Parent functional locations contain all of the attribute data applicable to each of their respective "child" functional locations.

Data Preparation

The objective of data preparation is to make PSOI information relational. This requires selection and configuration of a development platform that can be used as a repository for information and for creating interfaces between data from different sources. Microsoft Access is one application well-suited for this purpose due to its widespread distribution as a Microsoft Office product, its versatility in a wide range of applications and number of users, its scalability, and its security features.

PSOI is maintained in most facilities in a variety of media, which can be categorized as:

- Hardcopy files
- Non-relational electronic data such as scanned images and free-formatted text files
- Databases applications such as enterprise applications or single-user products

A comprehensive listing of PSOI that a stakeholder may want to make relational can be found in *Guidelines for Chemical Process Quantitative Risk Analysis* (CPQRA 2000, Section 5). Table 1 lists examples with recommended relationship types.

Table 1 – Examples of Relational PSOI Data

PSOI Data Table	***Relationship Type***	***Related Table***
Craft training/certification (equipment specific)	Many-to-many	Equipment
Maintenance procedures	Many-to-many	Functional location
Predictive and preventive maintenance (PPM)	Many-to-many	Equipment
Safe job procedures (SJP)	Many-to-many	Functional location
Spare parts	Many-to-many	Equipment
Incident reports	One-to-many	Functional location
Material safety data sheets (MSDS)	Many-to-many	Functional location
Production rates	One-to-one	Functional location
Risk scenarios	Many-to-many	Functional location
Work order data	Many-to-many	Equipment, Functional location

Hardcopy and non-relational files need to be reformatted where practical into relational database tables within the chosen development platform (referred to henceforth as PSOI application). Each table must have a primary key: a field or a combination of fields that holds a value that uniquely identifies each record. Tabular relationships are most commonly many-to-many and are established by use of junction tables. A junction table contains the fields that comprise the relationships and may also contain other fields for attributes specific to each record. The table "tblScenarioLink" in Figure 3 is an example of a junction table.

Data reformatting is tedious and time-consuming but nonetheless a very important step that needs to be done only once. It is important to note that while non-relational data files can be accessed from within a structured database via hyperlinks, the information within the files is of far less value to the user.

Some PSOI data will be maintained in other applications and will need to be interfaced with the PSOI application. One solution is to use routine data exports into an information warehouse, from which the PSOI application can upload and update or append data as required. MS Access does not allow users to enforce referential integrity with linked tables, so it is better to build tables within MS Access and append or replace data to make the tables current.

Example of Non-Relational versus Relational Data

A user is trying to identify all carcinogenic chemicals and where they are located in what quantities in his facility.

- ***Non-relational case*** – All MSDS data are stored as scanned images.
- ***Relational case*** – All MSDS data are in a database table that has a many-to-many relationship with a functional location table. The junction table linking the two has a field for chemicals quantity, which has been populated.

The ***non-relational case*** requires the user to review each MSDS one at a time to determine which chemicals contain carcinogens. He would then need to determine where each carcinogenic chemical is located in the facility and following that would need to determine the chemical quantities in each location. The time requirement would be many hours if not days, depending on the size of the facility.

The ***relational case*** requires the user to generate a database query. The time requirement would be less than five minutes and would be essentially independent of the size of the facility.

Figure 2 – The Power of Relational Data

Risk Assessment: the Cohesive Thread

Risk assessment is a process for compiling and reviewing applicable PSOI and using the data to define and tabulate potential scenarios for failure and estimate associated risk levels. If made relational, failure scenarios can give personnel accessible and concise summaries of hazards and operability issues and show where safeguards are applicable and which consequences they prevent. This in turn helps to prioritize safeguard support activities executed by operations and maintenance personnel and identifies where support activities are inadequate.

The PSOI application can be used to monitor the effectiveness of safeguards as they pertain to equipment reliability and adjust criticality rankings accordingly. It can then report increased or decreased criticality levels and the associated reasons to appropriate personnel for remedial action. The PSOI application can also give its users the capability to efficiently navigate through all scenarios from all methodologies from any starting point and display information such as:

- Scenarios from all methodologies in descending order of risk level for a given functional location.
- All initiating events related to specific equipment type failures in a facility, the unique ID of each equipment item involved in those events, and all safeguards in place to prevent those events.
- All consequences related to specified events.

Making Risk Scenarios Relational

The CPI use a wide variety of risk assessment methodologies to identify potential risk associated with the operation of their facilities. Table 2 lists some examples. While methodologies can vary significantly in complexity and type of approach, the end product is similar in that each produces a set of failure scenarios. A scenario includes at a minimum an initiating event and one consequence. It may also include subsequent events resulting from the initiating event (event tree), faults leading up to the initiating event (fault tree), identification of safeguards, additional consequences, and risk-ranking.

Each scenario has functional locations associated with it in one or more of the following roles: fault, event, or safeguard. Each associated functional location and its role(s) should be explicitly identified as being part of its parent scenario. The associations are built with many-to-many relationships. Figure 3 illustrates these relationships. Once the functional location associations have been made, all PSOI related to the functional locations will be linked to the parent scenario.

Table 2 – Examples of Risk Assessment Methodologies

- Equipment Criticality Ranking (ECR)
- Hazard and Operability (HAZOP) Studies
- Risk-based Inspection (RBI) Assessments
- Reliability-centered Maintenance (RCM) Assessments
- Pressure Relief System Assessments
- Failure Modes and Effects Analysis (FMEA)
- What-if & What-if/Checklist
- Layer of Protection Analysis (LOPA)
- Quantitative Risk Analysis (QRA)

Safeguards can be divided into equipment and non-equipment categories. Equipment safeguards are assets with equipment numbers that are installed in functional locations. Non-equipment safeguards such as procedures and inspections make equipment more reliable or otherwise reduce risk associated with scenarios. Typically only a subset of the total non-equipment safeguards linked to a given scenario via associated functional locations are applicable to that scenario. For example, a warehouse spare thrust bearing for Equipment No. 3021990, which is installed in functional equipment location YEM-CPF-TPL-TP1-DES-002 (Topping Plant Residue "A" Pump, Tag No. P-21441A), is a safeguard for this functional location, but does not reduce risk for Scenario 1758, which is built on the initiating event "The Topping Plant 1 Desalter Water Pump trips interrupting the flow of water to the Residue Pump seal flush." Each safeguard that is linked to a scenario and reduces risk should be associated via a many-to-many relationship with the respective consequence(s) and thereby the scenario for which risk is reduced.

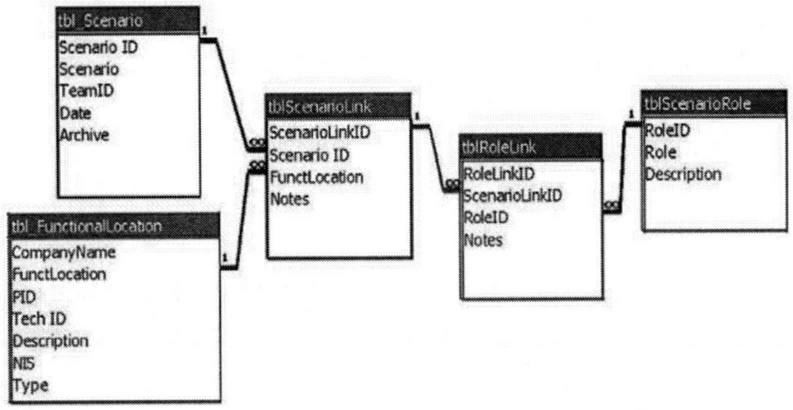

Figure 3 – Scenario to Functional Location Relationships

Real-time Risk Ranking

Risk assessment data sets are typically maintained as static documents. HAZOP studies, for example, are reviewed and updated once every five years per OSHA requirements. Risk assessment concerns are frequently alleviated with the implementation of safeguards, which in many cases are reliant on proper execution of inspections or procedures. If the inspections or procedures are not done or improperly executed, risk levels go back up, possibly even to pre-risk assessment amounts. The risk-based inspection recommended practice (API RP 580 2002, Section 12.1) states that "Risk mitigation achieved through inspection presumes that the organization will act on the results of the inspection in a timely manner. Risk mitigation is not achieved if inspection data that are gathered are not properly analyzed and acted upon where needed. The quality of the inspection data and the analysis or interpretation will greatly affect the level of risk mitigation. Proper inspection methods and data analysis tools are therefore critical." The same logic also applies to risk reduction measures other than inspections such as craft training/certification, spare parts inventory, and procedures.

Facilities need to be aware of fluctuations in risk levels so appropriate remedial action can be taken when required. The PSOI application can generate real-time risk data once numerical associations have been made between reliability, safeguard effectiveness, and safeguard support activities. The following discussion outlines this process.

Risk can be defined as a function of probability or frequency and consequence of a particular scenario with the following equation (CPQRA 2000, Section 5):

$$\text{Risk} = F(s, c, f) \qquad \text{(Equation 1)}$$

Where:
s = hypothetical scenario
c = estimated consequence(s)
f = estimated frequency

The estimated frequency of consequence occurrence is a function of the individual failure rates for equipment and safeguards that are part of a scenario. Process equipment failure rate data typically are generic and do not factor in a number of process influences. Generic data can be adjusted to account for such influences with the following equation (CPQRA 2000, Section 5.5.20):

$$\lambda_A = \lambda_g \prod_i^n f_i \qquad \text{(Equation 2)}$$

Where:
λ_A = adjusted equipment failure rate
λ_g = generic equipment failure rate
f_i = adjustment factor i (i = 1 to n)
n = number of adjustment factors that apply

CPQRA 2000 gives examples only of adjustments that increase failure rates, where $f_i \geq 1$, such as corrosion or vibration. Equation 2 can also be used to consider the increased reliability given by safeguards, where $f_i \leq 1$. Adjustment factors (f_i) for safeguards are in many cases arbitrary values derived from expert opinion or judgment and are based on assumed levels of safeguard effectiveness. When the level of effectiveness for a specific safeguard is not achieved, credit for that safeguard needs to be correspondingly reduced. For example, a relief valve has a probability of failure on demand (PFD) of 0.01 (LOPA 2001, Table 6-4). If the adjustment factor for timely execution of the PM inspection is conservatively set at $f = 0.02$ and one preventive maintenance testing interval for testing and inspection of a relief valve is not executed, the PFD for that valve increases to 0.05. The logic proposed by Equation 2 can also be used to make adjustments to consequence severity levels.

Risk reduction for installed equipment safeguards is treated in a similar manner. The Layer of Protection Analysis methodology (LOPA) shows that risk reduction for independent protection layers is represented by the following equation (LOPA 2001):

$$f_i^C = f_i^I \prod_{j=1}^{J} PFD_{ij} \qquad \text{(Equation 3)}$$

Where: f_i^C = the frequency for consequence C for initiating event i

f_i^I = the initiating event frequency for initiating event i

PFD_{ij} = is the probability of failure on demand of the ith IPL that protects against consequence C for initiating event i.

IPL = independent protection layers

The values of PFD_{ij} and f_i^I in Equation 3 are functions of maintenance inspections and other safeguards, which can be factored in by use of Equation 2.

A similar process for taking risk reduction can be used with qualitative risk analysis. A review team typically will take safeguard credits for a particular scenario and reduce the likelihood and/or consequence severity categories correspondingly. Credits are assigned for each safeguard as a multiplier of a risk category increment and may range in value from a fraction to a multiple of unity.

Normalizing Risk Ranking

Combining data from multiple risk ranking methodologies requires a standard unit of measure. Risk ranking is done using either qualitative or quantitatively methods or a combination of both. Qualitative results are generally in matrix form, displayed as consequence severity categories versus likelihood of occurrence categories. Risk matrices vary with different risk assessment and ranking methodologies, so it is not unusual for stakeholders to have multiple risk data sets presented in more than one matrix format. In many cases, the different matrix formats are neither comparable with each other nor with quantitative data.

A universal risk matrix can be used to represent both qualitative and quantitative data. Consequence and likelihood categories are assigned numerical value ranges, which form a grid onto which quantitative values are plotted. Data cross-reference tables are used to convert qualitative data in different matrix formats into the universal matrix format.

One difficulty with a universal matrix is assigning numerical values to health, safety, and environmental (hazard) consequence severity categories. This is best done by first developing consequence severity categories for process-related consequences, which are more-easily quantified, and subsequently developing hazard consequence severity categories by order-of-magnitude comparisons.

Classifications

Classifications make record navigation and selection more efficient. They allow selection and filtering of data by attributes, which speeds review and analyses. At a minimum, categories should be defined and used for equipment types and scenarios.

Conclusion

The CPI can benefit by using information management tools to organize their process safety and operability information. When information is organized and accessible, more people will use it and thus be better informed. Better information means better decisions. Better decisions mean safer facilities and more efficient operation.

References

Guidelines for Chemical Process Quantitative Risk Analysis, Second Edition, American Institute of Chemical Engineers, New York, NY, 2000.

Layer of Protection Analysis: Simplified Process Risk Assessment, American Institute of Chemical Engineers, New York, NY, 2001.

Risk-Based Inspection, API Recommended Practice 580 First Edition, American Petroleum Institute, Washington D.C., 2002.

A Novel Method Using Dynamic Time Warping Algorithm for Online Analysis of Batch Processes

NV. Saritha and Natarajan Venkateswaran
GE John F. Welch Technology Center, Bangalore, India.

Abstract:

Batch processes are affected by disturbances in process variables, errors in charging of recipe materials and existence of physical constraints. These can impact the quality of the product. Multivariate SPC charts may be applied to monitor batch-to-batch variations. These are based on analysis of data from past successful batches. Batches generally are not of equal duration. Hence to apply statistical analysis, one needs to synchronize batch trajectory data. This is conventionally achieved using dynamic time warping (DTW). DTW synchronizes two batch trajectories by appropriately translating, expanding, and contracting localized segments within both trajectories to achieve a minimum cumulated distance between the trajectories. However, it has some limitations since it focuses only on the magnitude differences. To overcome this, a method based on slope differences called Derivative dynamic time warping (DDTW) was proposed. This method also suffers from a few drawbacks.

This paper introduces a new technique, vector-based dynamic time warping (VDTW). The VDTW technique is an improvement over DTW and DDTW techniques. Being a vector-based approach, the new method considers the changes in magnitude and temporal derivative in the process variable. Examples are used to illustrate the performance of this method. Finally, a method to apply the technique for online implementation of statistical process monitoring is discussed.

1. Introduction

Batch processes are very common in chemical and pharmaceutical industries. Monitoring these processes is desirable for several reasons such as quality maintenance, quality improvement and safety. One of the methods for batch process monitoring is based on multivariate statistical techniques introduced by Nomikos and McGregor (1994). Batch processes in process industries are usually equipped with sensors that measure key process variables like temperature, pressure and flow rates. Monitoring these variables and inferring batch information like batch quality is the essence of statistical batch process monitoring.

For statistical batch process monitoring a database is needed with completed batch runs that produced on-spec products. The variation within this data serves as a reference distribution. The performance of new batches is then compared with this reference distribution, using multivariate control charts. These charts are normally based on principal component analysis (PCA) and Partial Least Squares (PLS) models. For some models it is required that the reference batches have equal duration. Usually batches are not of equal duration and even if it is so, similar events happen at different time points for different batches. In such cases, synchronization of events among batches is necessary. The method of dynamic time warping (DTW), proposed by Kassidas et al. (1998) is useful for synchronization of batches. Another recent method (Keogh and Pazzani, 2001) to do warping is the derivative dynamic time warping or DDTW. The DDTW method is based on slope comparison and overcomes the limitations of DTW for certain batch data.

In this work, DTW and DDTW techniques are discussed and are shown to be inadequate. A new warping technique based on vectorial comparison of batch trajectories is presented. The role of these warping techniques in online statistical batch process monitoring is discussed.

2.1. Dynamic Time Warping

Dynamic time warping, DTW, is a general method for warping trajectories with unequal duration and is based on dynamic programming.

Let the trajectory of a reference batch be denoted by a column vector $\boldsymbol{r}$ $(J \times 1)$ (where J is the number of points) and that of a sample batch by vector $\boldsymbol{s}(I \times 1)$. The objective is to make these two trajectories of equal length. The sample and the reference vectors, form the base of a two-dimensional grid as shown in Figure 1.

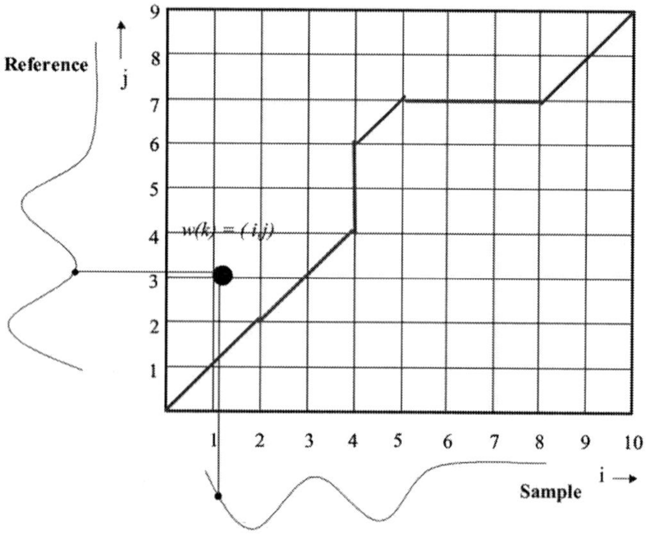

Figure 1: Warping Path

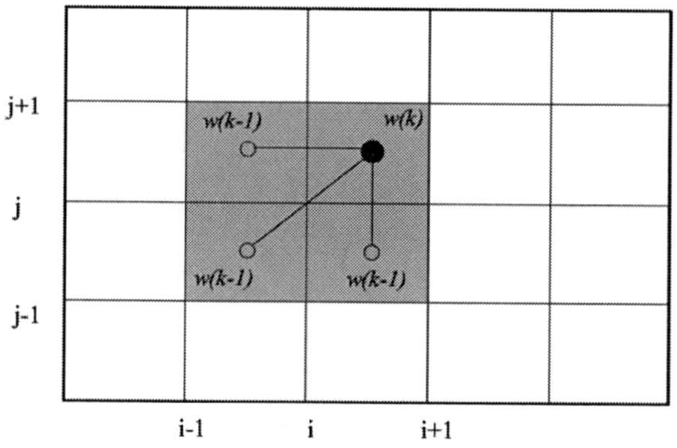

Figure 2: Local Constraints

The grid dimensions are the lengths of $\boldsymbol{r}$ and $\boldsymbol{s}$, for Y and X axes respectively. A grid point is denoted by $w(k)$ where k is an index. A sequence of grid points will be referred to as the trajectory or warping path $\boldsymbol{f}$.

$$\boldsymbol{f} = \{w(1), w(2), \ldots, w(k), w(k+1), \ldots, w(K)\}, \quad \max(I, J) \leq K \leq I + J \qquad (1)$$

where the number of grid points in the path is given by K. The starting point of $\boldsymbol{f}$ is $w(k = 1)$ which is given by $(i = 1, j = 1)$ and the ending point $w(k = K)$ by $(i = I, j = J)$. For any point along the path, $w(k)$ *is* denoted as $w(k) = (i(k), j(k))$. An example of such a path is given in Figure 1.

Many paths of different size *K*, can be constructed that connect the starting and end point of the grid.

The construction of a path ***f*** is also subject to local constraints. Consider the grid point *w*(*k*) illustrated as the black dot in Figure 2. This grid point (*i*, *j*) is part of trajectory ***f***. The predecessor *w*(*k* -1) of this grid point (*i*, *j*) can be any of the gray dots. That is:

$$w(k-1) = (i-1, j),\ (i-1, j-1) \text{ or } (i, j-1) \tag{2}$$

This guarantees that the path cannot go 'back in time' and therefore the slope between two consecutive grid points on the path is always ≥ 0.

2.2. DTW algorithm

DTW uses the principle of dynamic programming to minimize the dissimilarity measure (distance) between the two trajectories in order to map them. It non-linearly warps the two trajectories so that similar events are aligned thereby a minimum distance between them is obtained.

The cumulative distance *D*(*i*, *j*) from the first point *(1,1)* to point (*i*, *j*) along a path is given by:

$$D\,(i(k), j(k)) = \sum_k d(i(k), j(k)) = \sum_k d(w(k)) \tag{3}$$

where *d*(*w*(*k*)) is a distance measure between two points of the sample and reference. Commonly the squared Euclidean distance is used and is given by:

$$d(w(k)) = [s(i(k)) - r(j(k))]^2 \tag{4}$$

The goal of DTW is to find the optimal path ***f*** (that connects points (1,1) and (I, J)) with length *K* that minimizes the function *D*(*i*, *j*) in Eq. (3):

$$D(I, J) = \min_{(f)} \left[\, \sum_{k=1}^{K} d(w(k)) \,\right] \tag{5}$$

The optimal path is determined using dynamic programming in a back wise manner starting from the point *D* (*I*, *J*) as described by H. Sakoe, S. Chiba, (1978).

DTW has been successfully used in many domains. An example below illustrates how DTW correctly aligns the reference and sample batches. In Figure 3, the batch trajectories are plotted against data point number (time). Red lines join warped points across the two trajectories.

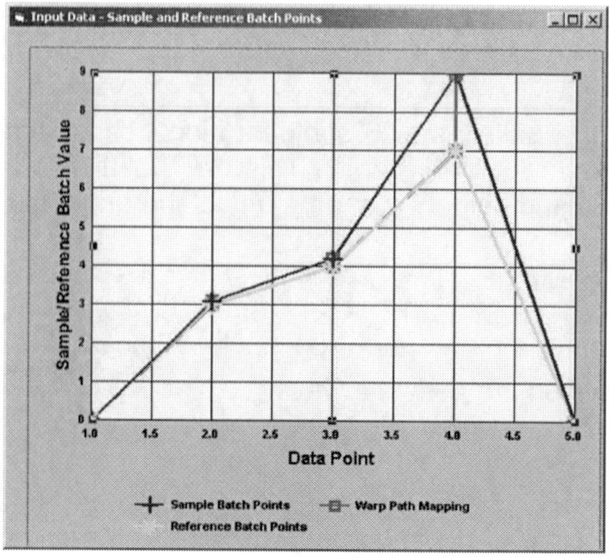

Figure 3: Case 1

3.1. Drawbacks of Euclidean Distance

DTW can lead to unintuitive alignments when a point on one trajectory warps on to more than one point on the other. This results in spurious warp points as explained in the following case.

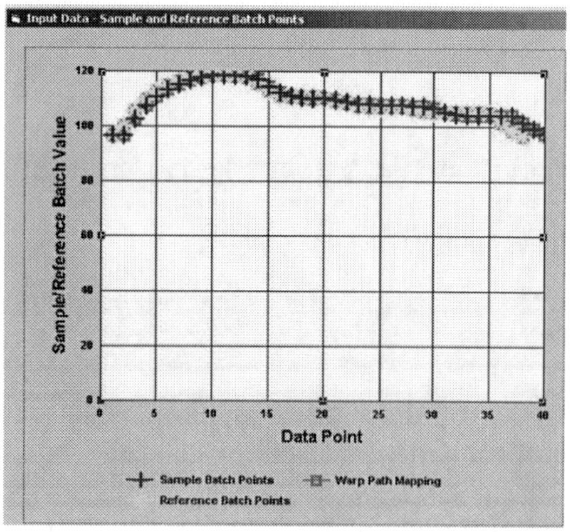

Figure 4: Case 2 - Reactor Temperature Vs Time

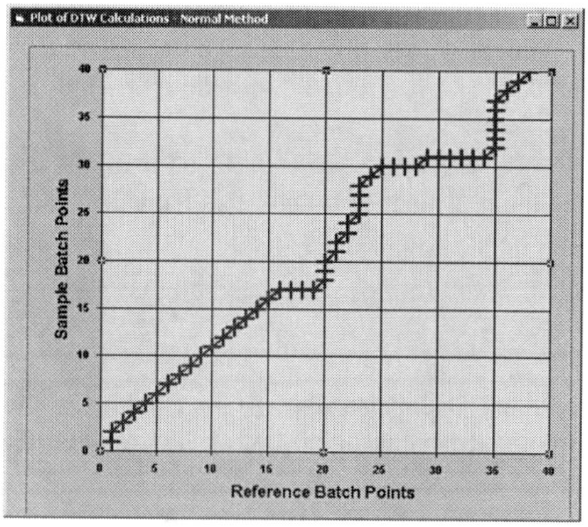

Figure 5: Case 2 - Warping Path using DTW

In Figure 4, a temperature profile for an industrial polymerization reactor is shown. The temperature varies in a narrow interval during the batch. Since the two batch profiles look similar we expect good alignment. Thus a point in a trajectory should be uniquely warped to a point on the other trajectory. This means that the warping path will lie close to the diagonal. However as explained earlier, the DTW technique will find spurious warp points.

In Figure 5, the warping path deviates from the diagonal. This effect is pronounced towards the later part of the batch (points15 and later) and comparison with Figure 4 shows that this occurs when the trend in temperature levels off.

Additionally, DTW algorithm may fail to find obvious, natural alignments in two sequences simply because a feature (i.e. peak, valley, inflection point, plateau etc.) in one sequence is slightly higher or lower than its corresponding feature in the other sequence (Keogh and Pazzani, 2001).

3.2. Derivative Dynamic Time Warping

As discussed above, DTW could map incorrectly points on a rising trend to a falling trend. To circumvent this problem, Keogh and Pazzani (2001) proposed the Derivative Dynamic Time Warping (DDTW) method, which considers the feature of "shape" and not the Y-values of the data points. The information on shape is obtained by considering the first derivative of the sequences to evaluate distances. The distance $d(w(k))$ between two points of the sample and reference is not Euclidean but rather the square of the difference in slope between points i and j.

$$d(w(k)) = \left[\frac{d(s(i(k))}{dt} - \frac{d(r(j(k))}{dt}\right]^2 \tag{6}$$

Applying DDTW technique for the same temperature example shown in Figure 5, we see that the warping path improves along the diagonal as depicted in Figure 6.

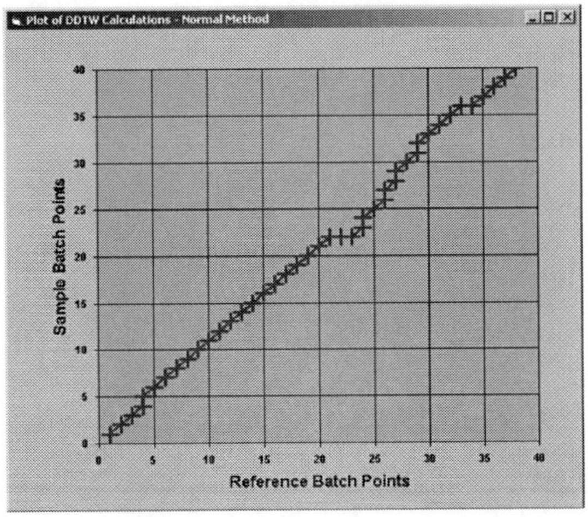

Figure 6: Warping Path for Case 2 using DDTW

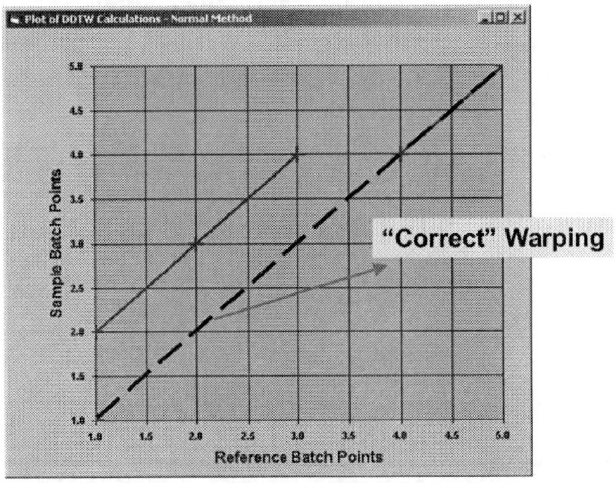

Figure 7: Warping Path for Case 1 using DDTW

However, DDTW doesn't account for the magnitude changes between two points. Applying DDTW to the data in Figure 3, it is evident that the technique does not correctly warp the trajectories (Figure 7). The dashed line represents the "correct" warping. We now introduce and demonstrate our method, which we call Vector-based Dynamic Time Warping (VDTW).

3.3 Vector-based dynamic time warping

The earlier techniques consider either the difference in magnitudes (DTW) or slopes between two points (DDTW) to compare similarity. To accomplish a better alignment we need to regard both the aspects and hence treat the points as "vectors" with attributes of magnitude and slope. Consider vectors Š(k) and •(k), of magnitudes s and r with slopes θ_1 and θ_2 respectively, as shown in Figure 8.

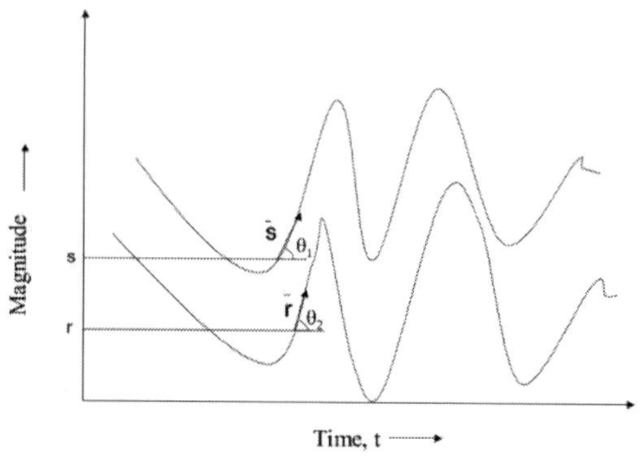

Figure 8: Vector representation of points for VDTW method

Our goal is to minimize the net accumulated vector distance D(*I*,*J*) as given in Eq. 5, where

$$d(w(k)) = || \check{S}(k) - \bullet(k) || \quad (7)$$

$$= || (s \cos\theta_1 - r \cos\theta_2)^2 + (s \sin\theta_1 - r \sin \theta_2)^2 ||^{1/2}$$

$$= || s^2 + r^2 - 2 s r \cos(\theta_1 - \theta_2) ||^{1/2} \quad (8)$$

The angles θ_1 and θ_2 are calculated using derivatives, which are evaluated as below.

$$\theta_1 = \tan^{-1} \left[\frac{d(s(i))}{dt} \right] \quad (9)$$

$$= \tan^{-1} \left\{ \frac{(s(i) - (s(i-1)) + \left[\{s(i+1) - s(i-1)\} / 2 \right]}{2} \right\} \quad (10)$$

This estimate of derivative is the average of two slopes viz. slope of the line through the point in question and its left neighbor, and the slope of the line through the left neighbor and the right neighbor. This estimate is more robust to outliers than any estimate considering only two data points (Keogh and Pazzani, 2001). For the first and last elements of the sequence we use the estimates of the second and penultimate elements respectively. Similarly θ_2 can be calculated using derivatives of •.

The DTW method minimizes the accumulated square of the distance (Eq. 4) whereas the DDTW method minimizes that of slope differences (Eq. 7), which are measures of dissimilarity between the two trajectories. The optimal path selection depends on this measure of dissimilarity, i.e., different measures can lead to different paths. Technically, it is the distance that needs to be minimized and not the square of the distance. So, we consider the vector distance between the two points. VDTW works successfully for both the cases shown in Figures 3 & 4.

Examples where the technique of VDTW works even when both DTW/DDTW fail are presented below. The robustness of VDTW is verified by considering two sine curves with magnitude variations and phase lag as in Figure 9. The warped points obtained using DTW technique are joined by lines. It can be seen that there are instances (highlighted in Figure 9) where a single point along the reference trajectory is warped to two adjacent points in the sample trajectory, which is incorrect.

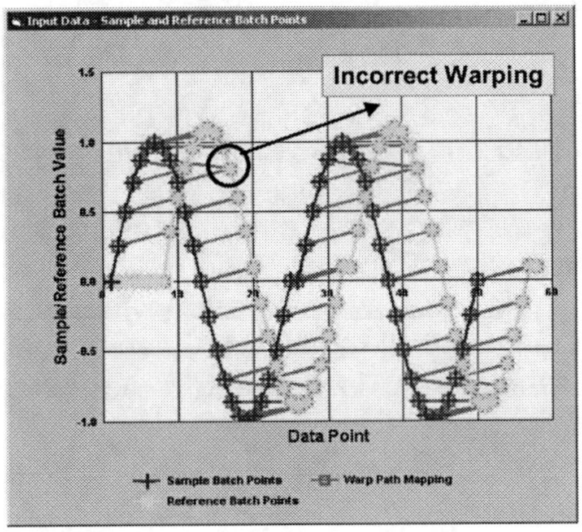

Figure 9: Case 3 – Mapping of Sine curves by DTW

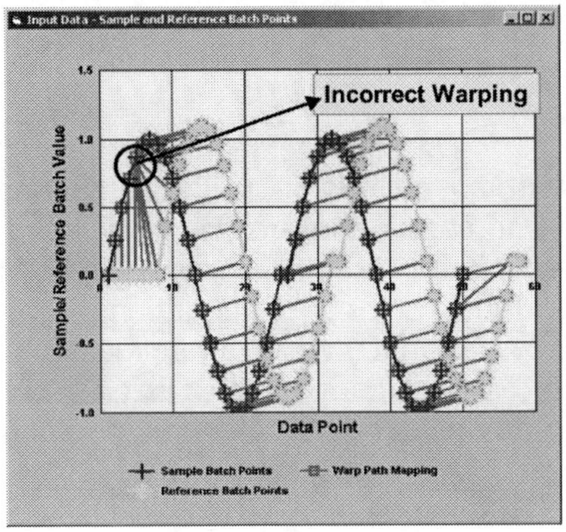

Figure 10: Case 3 – Mapping by DDTW

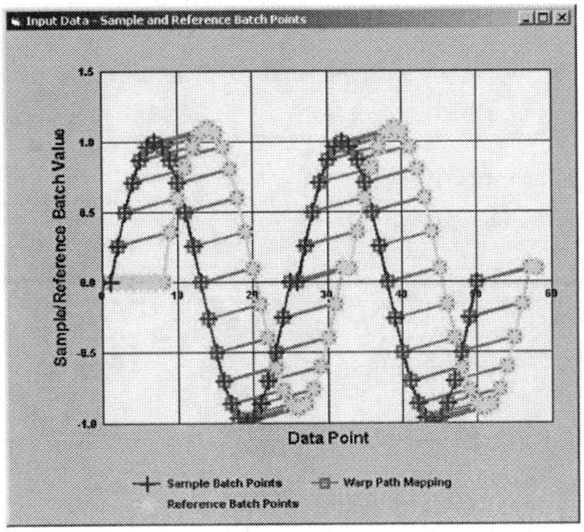

Figure 11: Case 3 – Mapping by VDTW

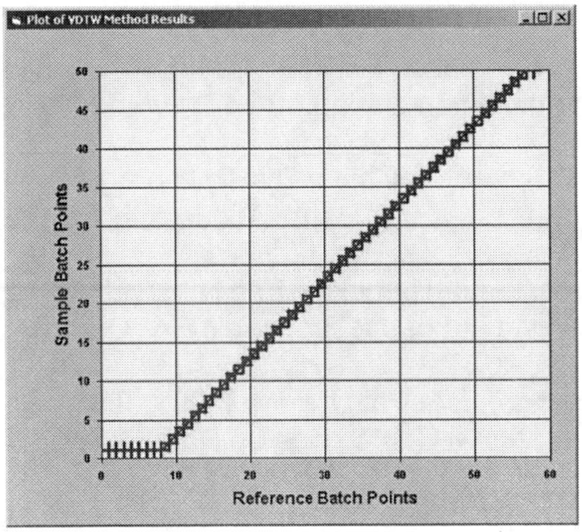

Figure 12: Case 3 – Warping Path by VDTW

In Figure 10, the same sine curves from Figure 9 are plotted but the warping in this case is done using the DDTW technique. There are a large number of points in the initial part (circled in Figure 10) of the actual trajectory, which are warped to the same reference trajectory point. When we apply the VDTW calculations to the same case, we see that the warping improves as shown in Figures 11 and 12.

4. Online Application

Online applications to monitor and quality control of batch processes in chemical process industries are an important and growing area. The warping techniques like DTW, DDTW and VDTW discussed in this paper have an important role in online implementation of the statistical batch process monitoring models. The online applications need to deal with synchronizing batches of unequal duration and this also includes batches that are still in progress. Guidelines for online implementation of DTW technique is given by Nomikos and MacGregor (1994) and similar methodology could be applied for VDTW technique.

5. Conclusions

Warping techniques, notably the dynamic time warping and the derivative dynamic time warping are available for comparison of batch information. These techniques work well for some cases and are shown to have limitations in correctly warping some other cases. A new technique, Vector Dynamic Time Warping (VDTW) is proposed. This technique is based on correct (vectorial) comparison of batch trajectories and can do better warping even when both DTW and DDTW are unsuccessful. Applications for online statistical monitoring have a need for batches to be synchronized and the warping techniques are useful in this regard. A method to do online statistical monitoring of batch processes is outlined.

6. References

1. Nomikos, P., & MacGregor, J. F. (1994). Monitoring batch processes using multiway principal component analysis. AIChE J. 40 (8), 1361_ 1375.

2. A. Kassidas, J.F. MacGregor, P.A. Taylor, Synchronization of batch trajectories using dynamic time warping, AIChE J. 44 (1998) 864–875.

3. H. Sakoe, S. Chiba, Dynamic programming algorithm optimization for spoken word recognition, IEEE Trans. Acoust. Speech Signal Proc. 26 (1978) 43–49.

4. Henk-Jan Ramaker, Eric N.M. van Sprang, Johan A. Westerhuis, Age K. Smilde, Dynamic time warping of spectroscopic BATCH data, Analytica Chimica Acta, 498 (2003) 133–153

5. Keogh, E. & Pazzani, M. (2001). Derivative Dynamic Time Warping. In First SIAM International Conference on Data Mining (SDM'2001), Chicago, USA.

7. Acknowledgments

We thank General Electric Advanced Materials for all the support and encouragement throughout the course of this project.

On the Use of Cutrices in Restriction Mapping

Dr. Kal Renganathan Sharma PE
Formerly Principal
Anna University affiliated, Sakthi Engineering College
Mount Poonamalle Road, Ramapuram (opp. MGR Gardens)
Chennai, Tamil Nadu 613402

Research Projects:

- New Data Structures in Sequence Analysis.
- Multi-Dimensional Matrices.
- Finite Speed Diffusion in Gel Electrophoresis.
- Reactions in Circle.

1

New Data Structures in Sequence Analysis

- Binomial Tree – Maximum Subseqence problem.
- Suffix Tree – Streaming a sequence – MUMMER.
- PQ Tree – STS Mapping problem with Perfect Data.

Maximum Subsequence Problem

- For example S={11,17,5,8,6,4,7,12,3}
- Recursion Solution
  - Given max. subsequence of Sn-1, maximum subsequence of Sn where n- number of elements can be found.
  - Needs right S1 for O(n) solution –Practical purposes $O(n^2)$ time taken.

Binomial Heap

- (Cormen,Leiserson,Rivest,Stein), Algorithms , MIT Press, Second Edition,2001.
- A Binomial Heap H is set of binomial trees that satisfies the following binomial heap properties.

  →Each Binomial tree in H obeys the minimum heap property: the key of a node is greater than or equal to the key of its parent. Such a tree is minimum heap ordered.

  →For any non-negative integer k, there is most one binomial tree in H whose root degree k.

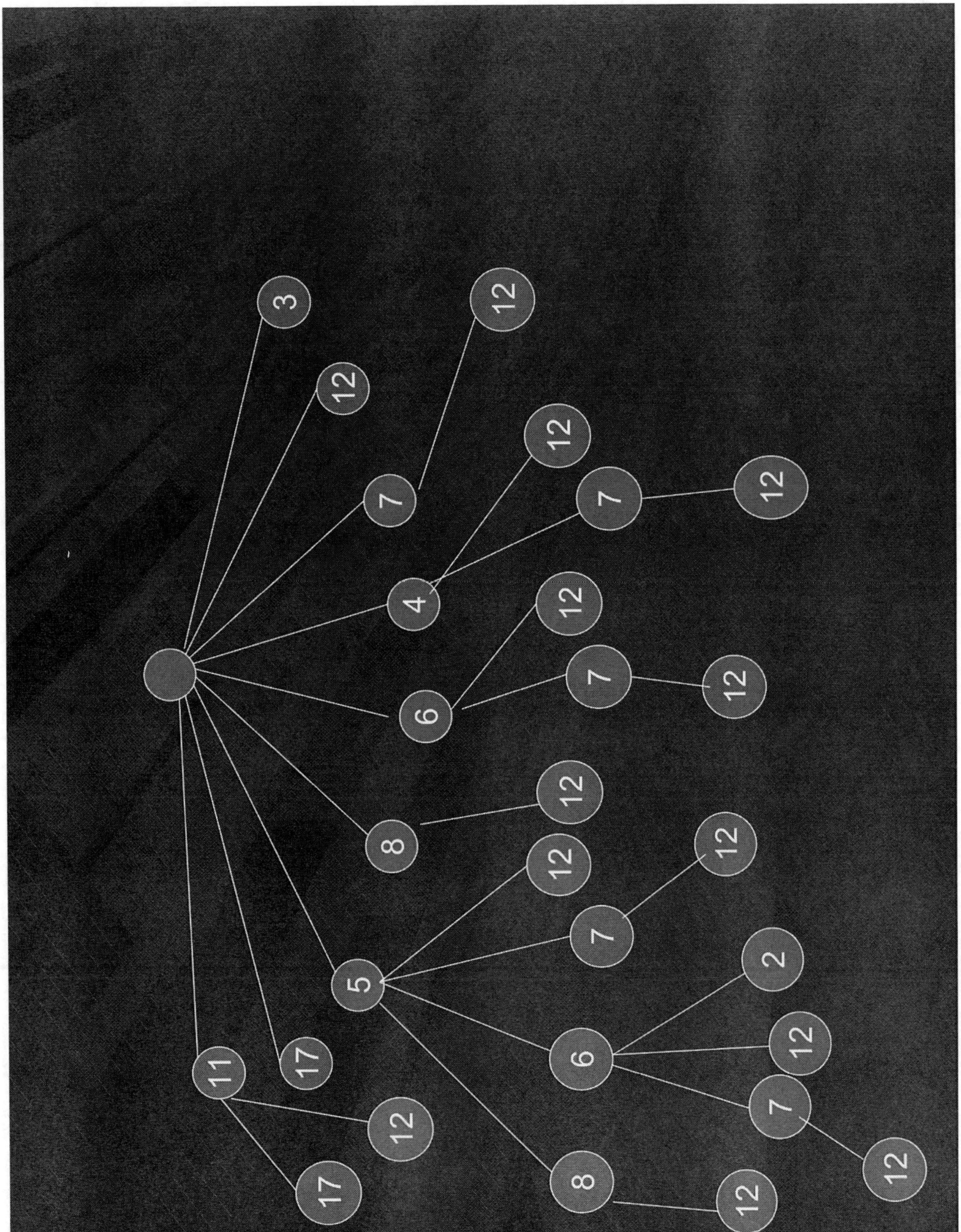
3
12
12
7
12
7
12
4
12
7
12
6
12
8
12
12
7
2
5
12
6
7
11
17
12
17
8
12
12

Binomial Heap

(K.R.Sharma ,2003, Binomial Tree Representation for the Maximum Subsequence Problem.
Norhtwest Regional Meeting of ACS, Bozeman,MT, USA)

- Max Subsequence –Deepest Branch
- {5,6,7,12}
- Time taken O(n).

Suffix tree representation of a sequence

- Sequence : TATATAAGGCGCGTT
  - 1 2 3 4 5 6 7 8 9 101112131415

Position	Longest sequence without repeat
1	TATAT
2	ATAT
3	TATAA
4	ATAA
5	TAA
6	AA
7	AG
8	GG
9	GCGC
10	CGC
11	GCGT
12	CGT
13	GT
14	TT

- Suffix Tree Representation of S=TATATAAGGCGCGTT

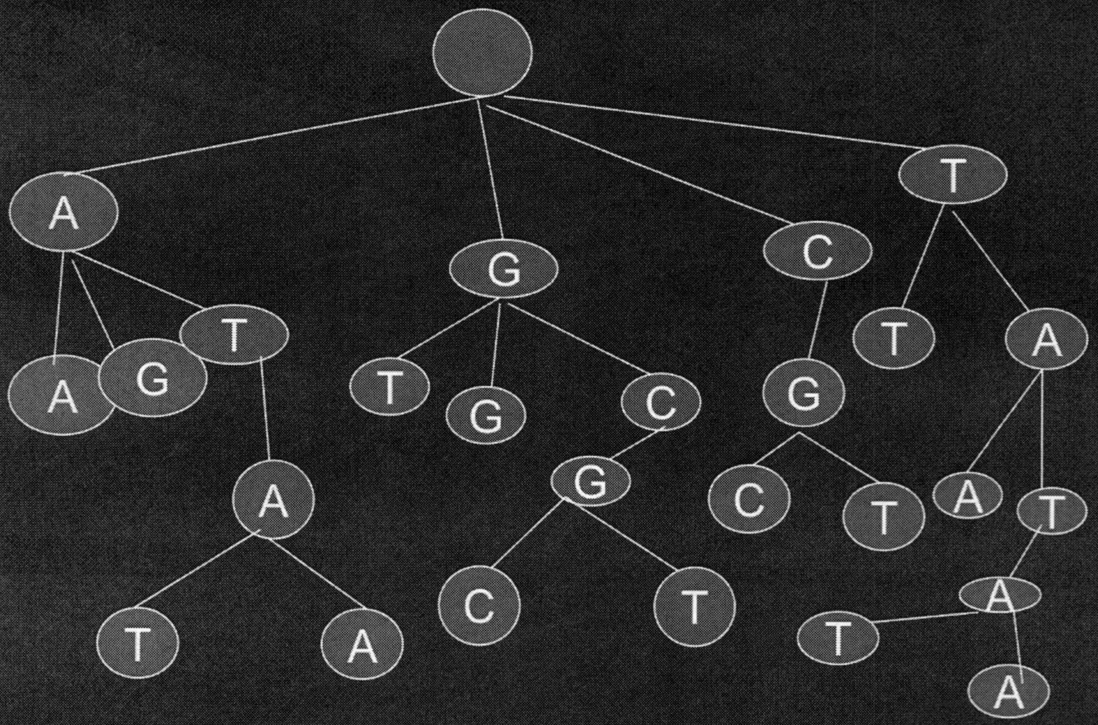

Cutrices in incidence relations

- Restriction Mapping
- Binary case
  - I(A,B)=I(A,AΛB) I Transpose(B,A ΛB)
  - Waterman,1995
  - 2 enzymes

- Suffix Trees of Two Sequences
- Position wisw Comparison of Repeats
- Global Alignment in O(n.cm^2) = O(n)
- Cm is the average length of Repeat
- Near Optimal Alignment
- Potential for Multiple Alignment

- I(A,A ΛB)

	1	2	3	4	5	6
1	0	0	0	0	0	0
2	0	1	1	0	0	0
3	0	0	0	1	0	0
4	0	0	0	0	1	1

- I(B, A ΛB)

	1	2	3	4	5	6
1	0	0	0	0	0	1
2	1	1	0	1	1	0
3	0	0	1	0	0	0

- I(A,A ΛB) I transpose(B,A ΛB)

$$\begin{pmatrix} 1 & 0 & 0 & 0 & 0 & 0 \\ 0 & 1 & 1 & 0 & 0 & 0 \\ 0 & 0 & 0 & 1 & 0 & 0 \\ 0 & 0 & 0 & 0 & 1 & 1 \end{pmatrix} \quad \begin{pmatrix} 0 & 1 & 0 \\ 0 & 1 & 0 \\ 0 & 0 & 1 \\ 0 & 1 & 0 \\ 0 & 1 & 0 \\ 1 & 0 & 0 \end{pmatrix}$$

- I (A,B)=

	1	2	3
1	0	1	0
2	0	1	1
3	0	1	0
4	1	1	0

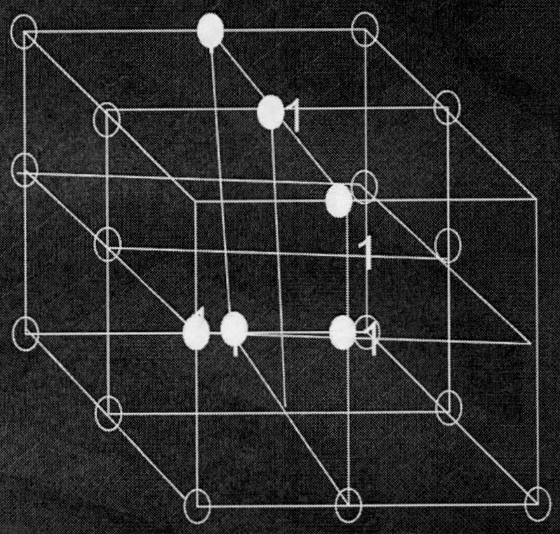

3X3X3 = 27

I(1,2,3) = 1   I(1,2,1) = 1

I(1,2,2) = 1   I(2,2,1) = 1

I(2,1,1) = 1   I(3,2,2) = 1

I(3,3,2) = 1

rows columns, floors

3 dimensional matrix – c

cutrice multiplication, Transpose Rules can be written

Xil = I(A,AΛBΛC)   Yjl = I(B,AΛBΛC)

Zkl = I(C,AΛBΛC) $\sum_{e}$XilYjlZkl

I(A,B,C) = I(A,AΛBΛC) [I(B, AΛBΛC), IT(C, AΛBΛC)

n X m+n+0   o X m+n+0

3X3X3 = 27

I(1,2,3) = 1 I(1,2,1) = 1

I(1,2,2) = 1 I(2,2,1) = 1

I(2,1,1) = 1 I(3,2,2) = 1

I(3,3,2) = 1

Rows, Columns, Floors

3 dimensional matrix – c

Cutrice Multiplication,

New Transpose Rules can be written

$X_{il} = I(A, A \wedge B \wedge C)$ $Y_{jl} = I(B, A \wedge B \wedge C)$

$Z_{kl} = I(C, A \wedge B \wedge C)$ $\sum_{e} X_{il} Y_{jl} Z_{kl}$

I(A,B,C) =

$I(A, A \wedge B \wedge C)\ [I(B, A \wedge B \wedge C), I^{T}(C, A \wedge B \wedge C)]$

n X m+n+0 o X m+n+0

PQ Tree Algorithm – STS Mapping Problem(perfect Data) (R.M.Karp)

- PQ Tree is a Data Structure
- Represents a certain class of permutations
- M Matrix Row – for each STS

  Column – for each clone M(i,j) = X or

  1,ifSTS in row I occurs on clone j
- P- node is one whose children may be re-arranged in any order
- Q-node is one whose children remain consecutive but may be arranged

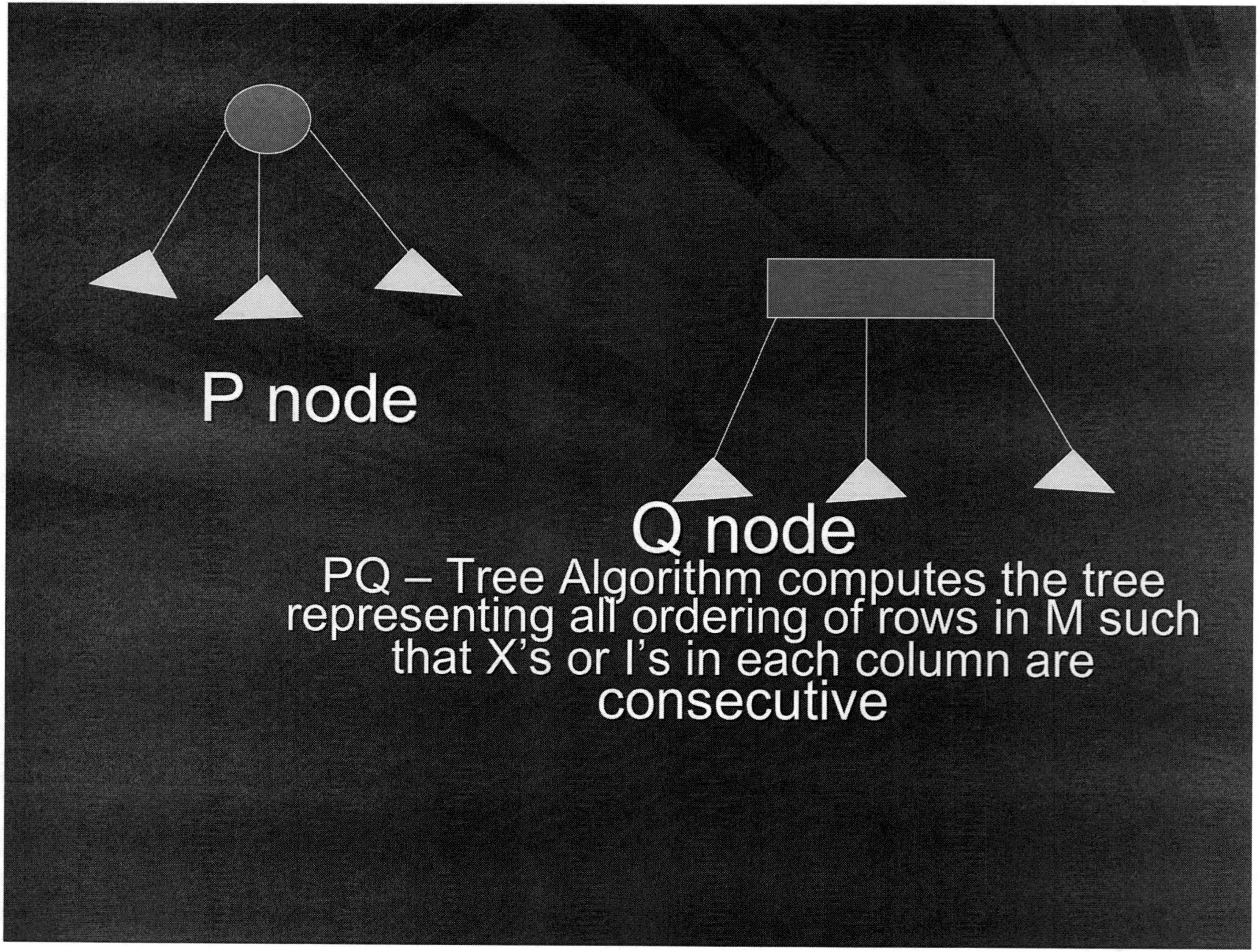
P node
Q node
PQ – Tree Algorithm computes the tree representing all ordering of rows in M such that X's or I's in each column are consecutive

M matrix for exmple – let A-G be 7 STS's on a DNA segment that is covered by 6 overlapping clones

	1	2	3	4	5	6
A	0	0	1	1	1	0
B	1	1	0	0	0	0
C	0	0	1	1	1	0
D	0	0	0	0	0	1
E	1	0	0	0	0	0
F	0	1	1	0	1	0
G	0	0	1	0	0	1

Rearrange rows in the above matrix so that they occur in the same order as the STS's occur then in each column all the is are next to each other

	1	2	3	4	5	6
E	1	0	0	0	0	0
B	1	1	0	0	0	0
F	0	1	1	0	1	0
A	0	0	1	1	1	0
C	0	0	1	1	1	0
G	0	0	1	0	0	1
D	0	0	0	0	0	1

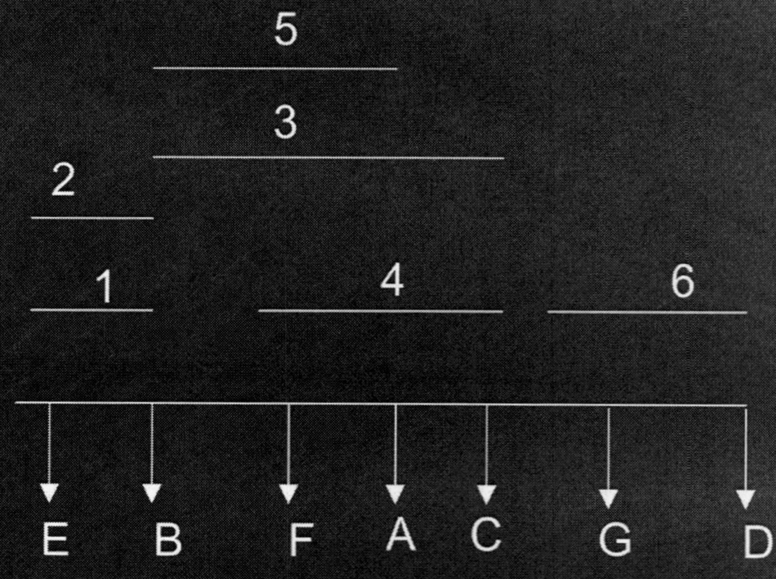

- At each iteration, a column is selected,which places a contraint on some subset of the leaves.

- Algorithm proceeds, column by column, until all constraints have been added.
- Relations in linear time

  4 ordership read of final tree --------->

  EBFACGD, DGCAFBE

  EBFCAGD DGACFBE

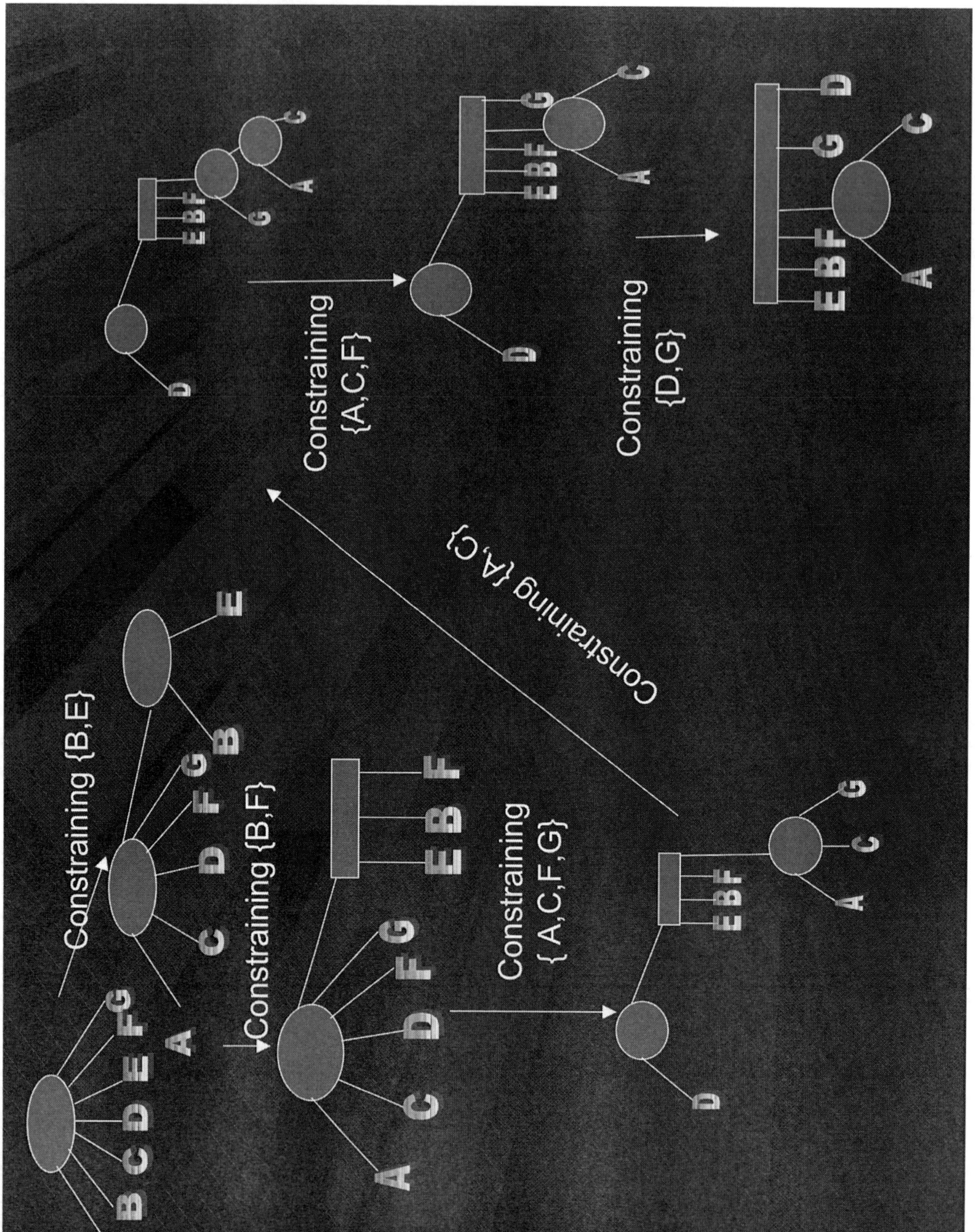
Constraining {B,E}
Constraining {B,F}
Constraining { A,C,F,G}
Constraining {A,C}
Constraining {A,C,F}
Constraining {D,G}

Pattern Recognition, Dynamic Regression, and Digital Filter in Chemical Engineering Education

Patrick Lo[1], Bill Swan[1], Daniel H. Chen[1], Patrick Wetuski[2], and Jack Stout[2]

Lamar University, Beaumont, Texas[1]

Nexus Engineering, Kingwood, Texas[2]

AIChE Spring National Meeting, New Orleans, LA
April 26-29, 2004

Outline of Contents

- Pattern Recognition
- Digital Filter
  - Optimal exponential moving average (EMA) filter constant for minimum time delay & maximum smoothness & fidelity
- Parameter identification with dynamic regression

NEXUS ENGINEERING

Chen, Lo, Swan, Wetuski, & Stout

Pattern Recognition (1)

- Process Monitoring/Diagnostics—One unexpected downtime can cost the plant a year's worth of good control
- Pattern recognition for preventive maintenance

NEXUS ENGINEERING

Chen, Lo, Swan, Wetuski, & Stout

Pattern Recognition (2)

- Pattern recognition was performed with statistical tools
  - number of times crossing zeros
  - unusual spikes
  - frequency analysis, i.e., Fast Fourier Transform (FFT)

NEXUS ENGINEERING

Chen, Lo, Swan, Wetuski, & Stout

Detecting the frequency of oscillation

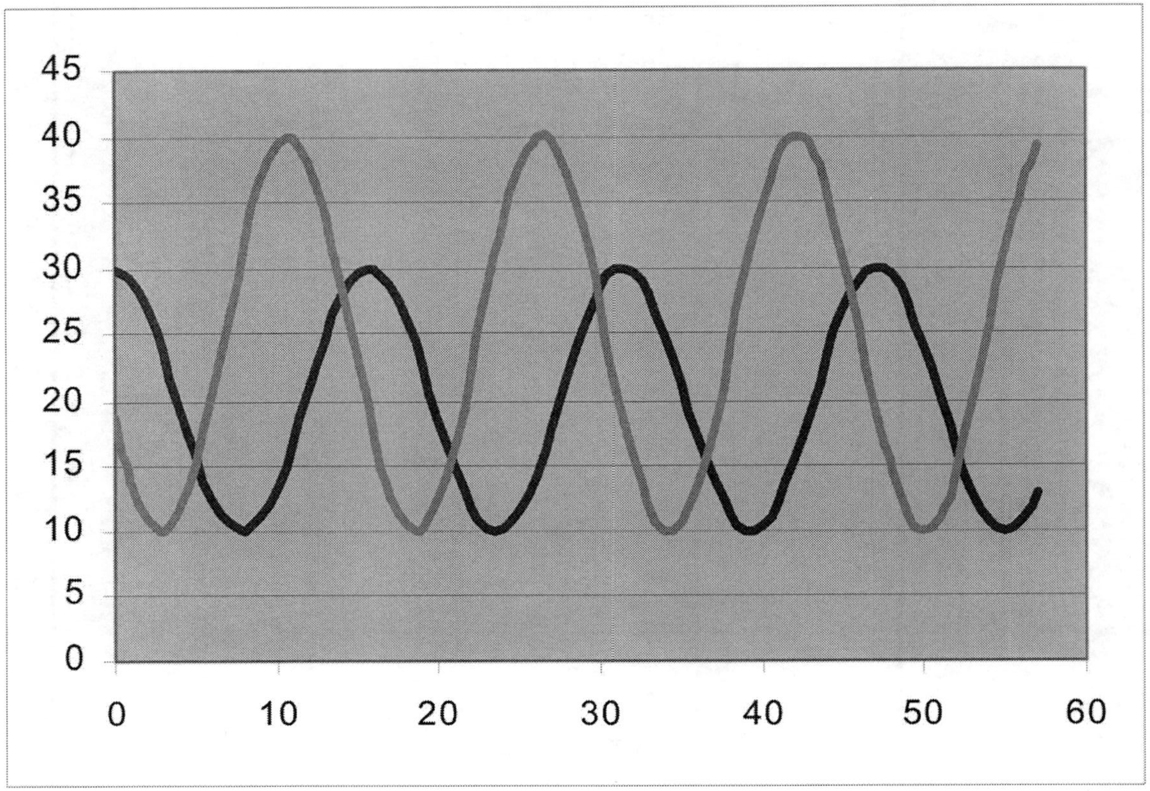

Controller PVs: Time Series

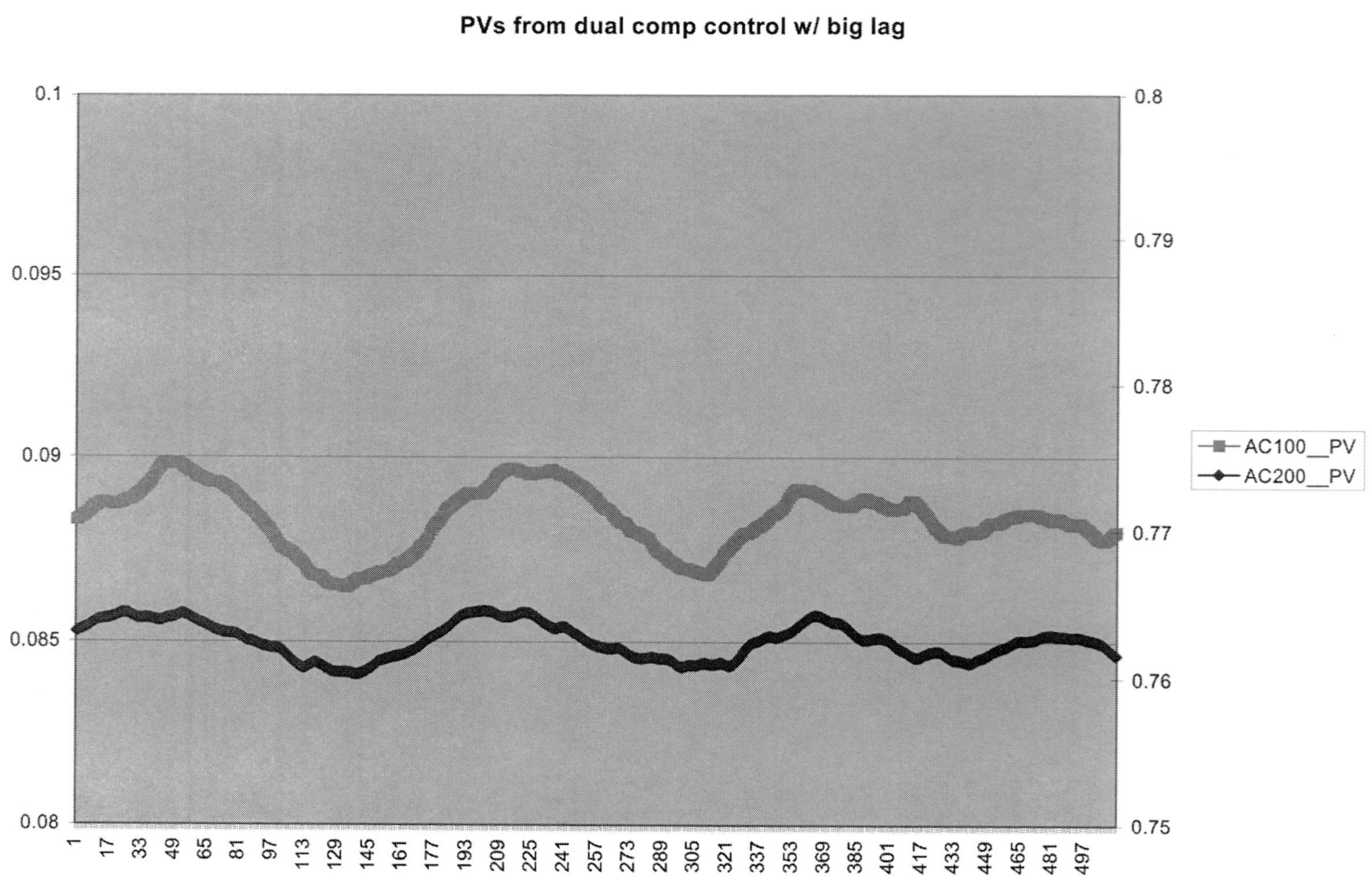

Fast Fourier Transform

- Sampling Frequency Must Be at Least Twice Signal Frequency to Detect Signal, e.g.,
  - signal frequency =2 cycles/min;
  - sampling frequency ≥ 4 samples/min.
- Requirements
  - 2^n Samples, where n is an Integer >2
- Results: Complex Numbers (x+yi)
- Number of Complex Numbers = # of Samples

NEXUS ENGINEERING

Chen, Lo, Swan, Wetuski, & Stout

Discrete Fourier Transform (1)

- Continuous Fourier Transform

$$X(\omega) = \int_{-\infty}^{\infty} x(t)\, e^{-j\omega t}\, dt$$

- Discrete Fourier Transform

$$X(k\Delta\omega) = (1/N)\sum_{n=0}^{N-1} x(n\Delta t)\, e^{-j(2\pi nk/N)}$$

where n= 0,1,2,…, N
k=0,1,2,….., N
N= # of samples

Discrete Fourier Transform (2)

- Identity

$$e^{-j(2\pi nk/N)} = \text{Cos}\,(2\pi nk/N) - j\,\text{Sin}\,(2\pi nk/N)$$

- Frequency Resolution

$$\Delta\omega = 2\pi/N$$

- Discrete Fourier Transform

$$X(k) = (1/N)\sum_{n=0}^{N-1} x(n)\, e^{-j(2\pi nk/N)}$$

Controller PVs: FFT Frequency analysis

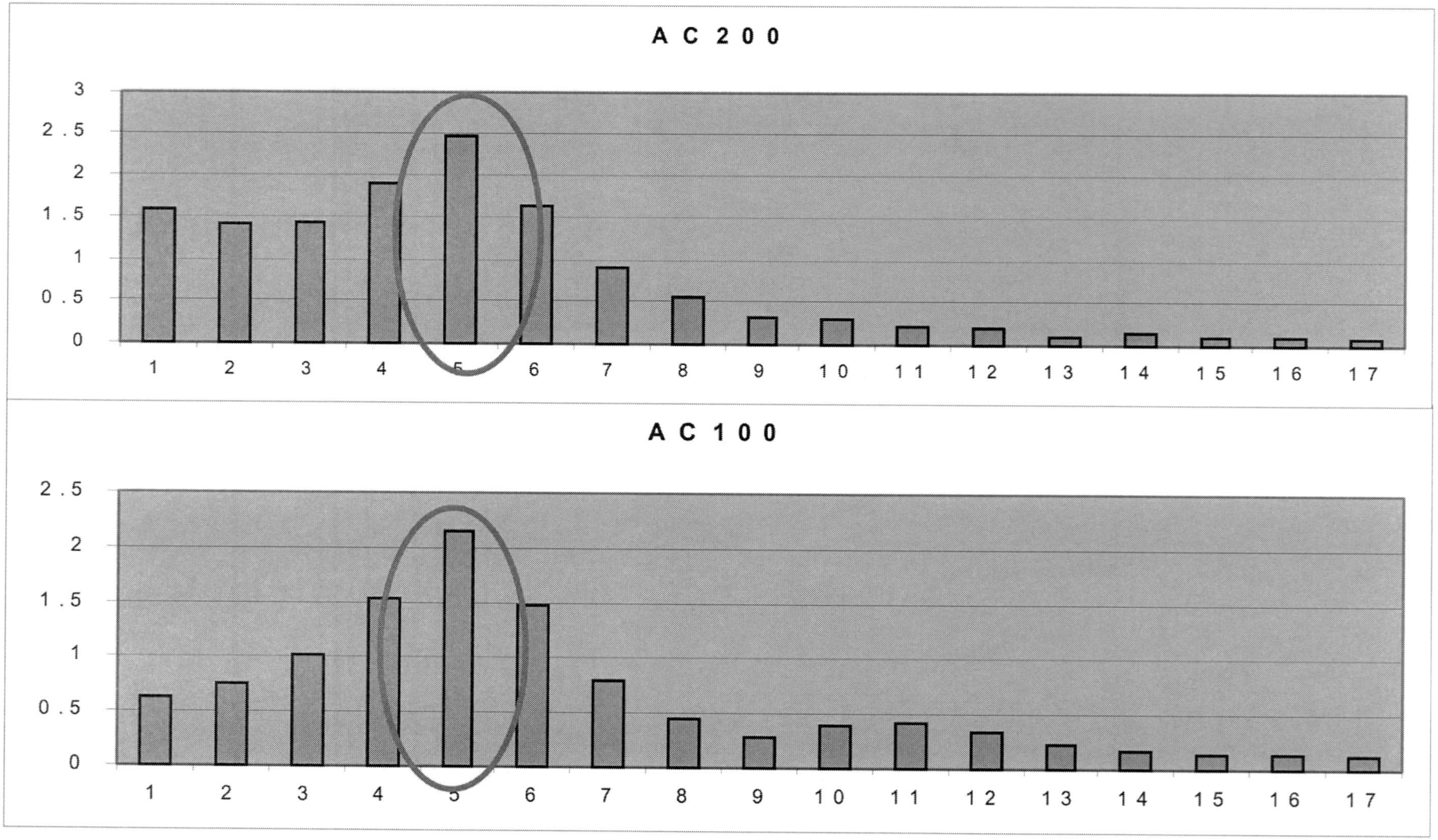

Digital Filters for Noisy Signal Processing

Chen, Lo, Swan, Wetuski, & Stout

NEXUS ENGINEERING

Noisy Signals

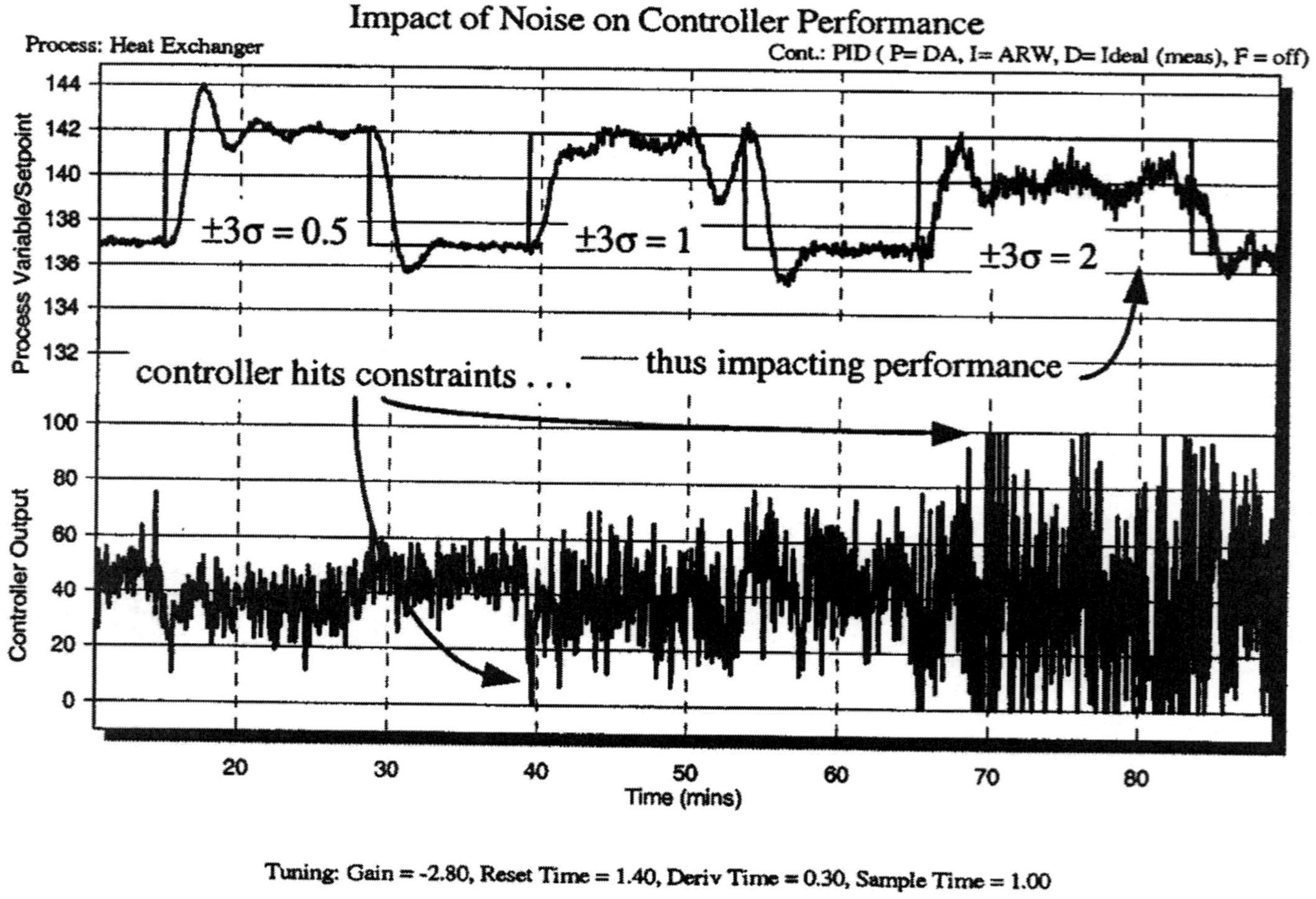

Control Station Manual, D. Cooper (2003)

Chen, Lo, Swan, Wetuski, & Stout

Continuous vs. Discrete Exponential Moving Average (EMA)

- $\tau_f * dz/dt + z = y$

  Where τ_f = first-order filter time constant

  y = unfiltered sensor reading

  Z = filtered reading

- Take first-order backward finite difference & Rearrange

  $Z_t = f * y_t + (1-f) * Z_{t-1}$

  Z_t = EMA, smoothed value

  y_t = unfiltered raw data

Chen, Lo, Swan, Wetuski, & Stout

Exponential Moving Average (EMA)

- EMA is a low-pass filter given by

$$Z_t = f * y_t + (1-f) * Z_{t-1}$$
$$= f * (y_t + (1-f) * y_{t-1} + (1-f)^2 * y_{t-2} + \ldots)$$

- f can be viewed as the current weight; and (1-f) as the memory
- The earlier data were given successively lower weights (1-f), (1-f)^2, (1-f)^3,….
- therefore the namesake "exponentially weighted" moving average or "exponential MA"

Chen, Lo, Swan, Wetuski, & Stout

NEXUS ENGINEERING

Example: EMA

- Data

  y_t , – 1, 2, – 3, 1, – 2

  $Z_0 = 0$, $f = 0.5$

  $Z_t = f * y_t + (1 - f) * Z_{t-1}$

- Statistics

  $Z_1 = 0.5^* (-1) = -0.5$

  $Z_2 = 0.5^*2 + 0.5^* (-0.5) = 0.75$

  $Z_3 = 0.5^*(-3) + 0.5^* (0.75) = -1.125$

  $Z_4 = 0.5^*(1) + 0.5^* (-1.125) = -0.0625$

  $Z_5 = 0.5^*(-2) + 0.5^* (-0.0625) = -1.03125$

NEXUS ENGINEERING

Chen, Lo, Swan, Wetuski, & Stout

Problem: Phase Delays

- These widely used digital filters are always accompanied with delays!
- The delays hamper control and make "in-time" inference and decision- making difficult.

NEXUS ENGINEERING

Chen, Lo, Swan, Wetuski, & Stout

Nasdaq Index vs. 50-day & 200-day Exponential Moving Averages

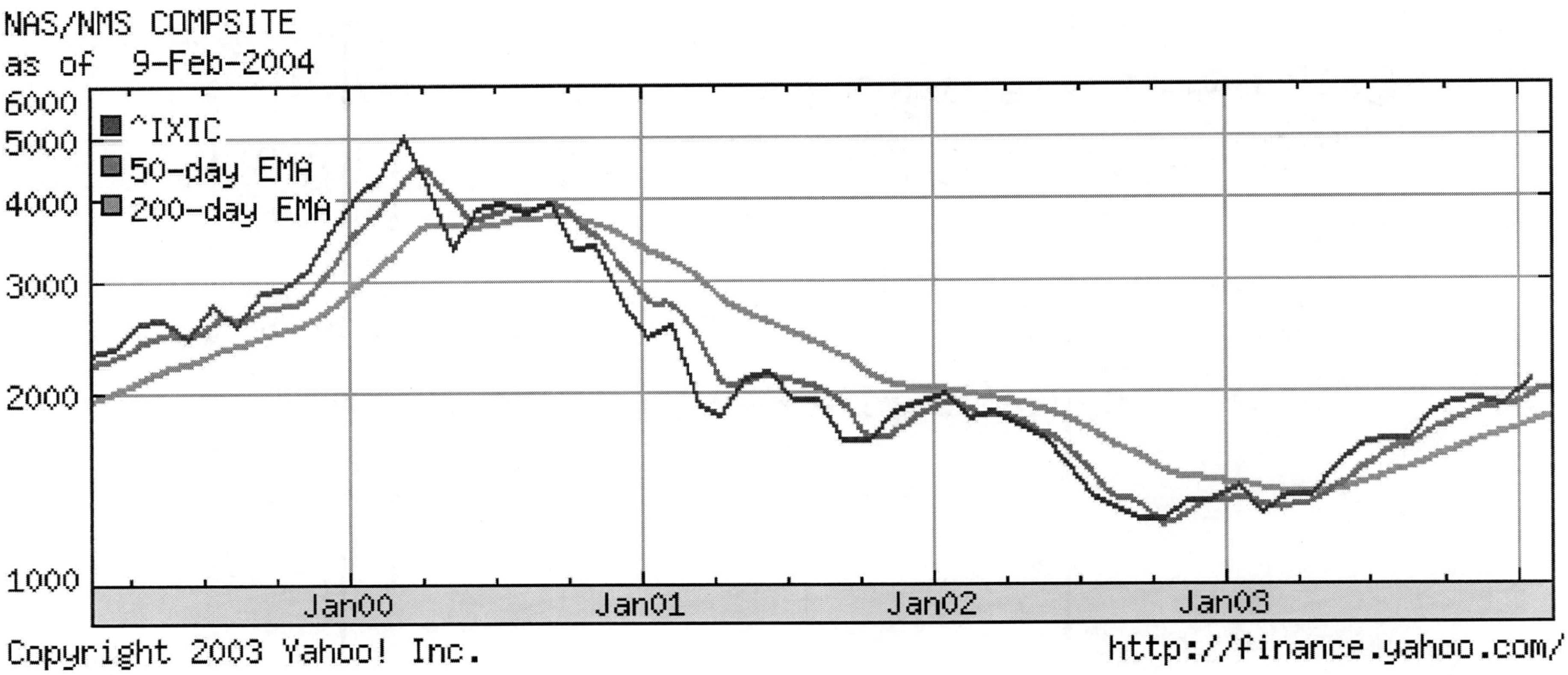

Lags

- For MA, the lag is half of the time window. For example, in 50 day MA the lag is 25.

- For EMA,

  $Lag = \tau_f / \Delta t$

  τ_f : time constant of the first-order filter

  Δt : sampling time.

- Also for EMA, filter constant $f = 1/(1 + Lag)$; $Lag=(1/f)-1$. For example, if $\tau_f = 1$, $\Delta t = 1$, lag = 1 , and f = 0.5.

Phase Lag

- The phase lag is equal to

θ = Lag* 2π /T

= Lag* 360°/T

Lag : time lag

T : period of the cycle T

Chen, Lo, Swan, Wetuski, & Stout

Criteria for Choosing Optimum Filter Constant in EMA

- EMA filter is a low-pass filter which removes high frequency noises at the expenses of a phase delay.

- Normally, an "eyeball" approach is taken to pick up a filter constant f that gives a "smooth" curve with a "tolerable" delay

NEXUS ENGINEERING

Chen, Lo, Swan, Wetuski, & Stout

Criteria for Choosing Optimum Filter Constant in EMA (Cont.)

- Definition Noisiness=Normalized Sum of Second Derivatives

  Smoothness = 1 – Noisiness
- Goodness of fit is represented by R-Squared between EMA and the original data. This is the fidelity to the original data.

Optimum EMA (1)

- Minimize delay yet to maximize goodness of fit and smoothness
- Max f*R^2*Sm

  f = EMA filter constant

  sm=smoothness=1-noisiness

  (Since Lag=(1/f)-1, max. f will min. lag)
- @ maximum of f*R^2*Sm, f = 0.59, this f emphasizes a low delay.

NEXUS ENGINEERING

Chen, Lo, Swan, Wetuski, & Stout

Optimum EMA (2)

- Correlation of coefficient R also represents the quality of fitting
- Max f*R*Sm

  f = EMA filter constant

  sm=smoothness=1-noisiness

  Lag=(1/f)-1
- @ maximum of f*R*Sm, f = 0.57, this f emphasizes a low delay.

NEXUS ENGINEERING

Chen, Lo, Swan, Wetuski, & Stout

Trade-Off Among Goodness of Fit, Smoothness, and Delay

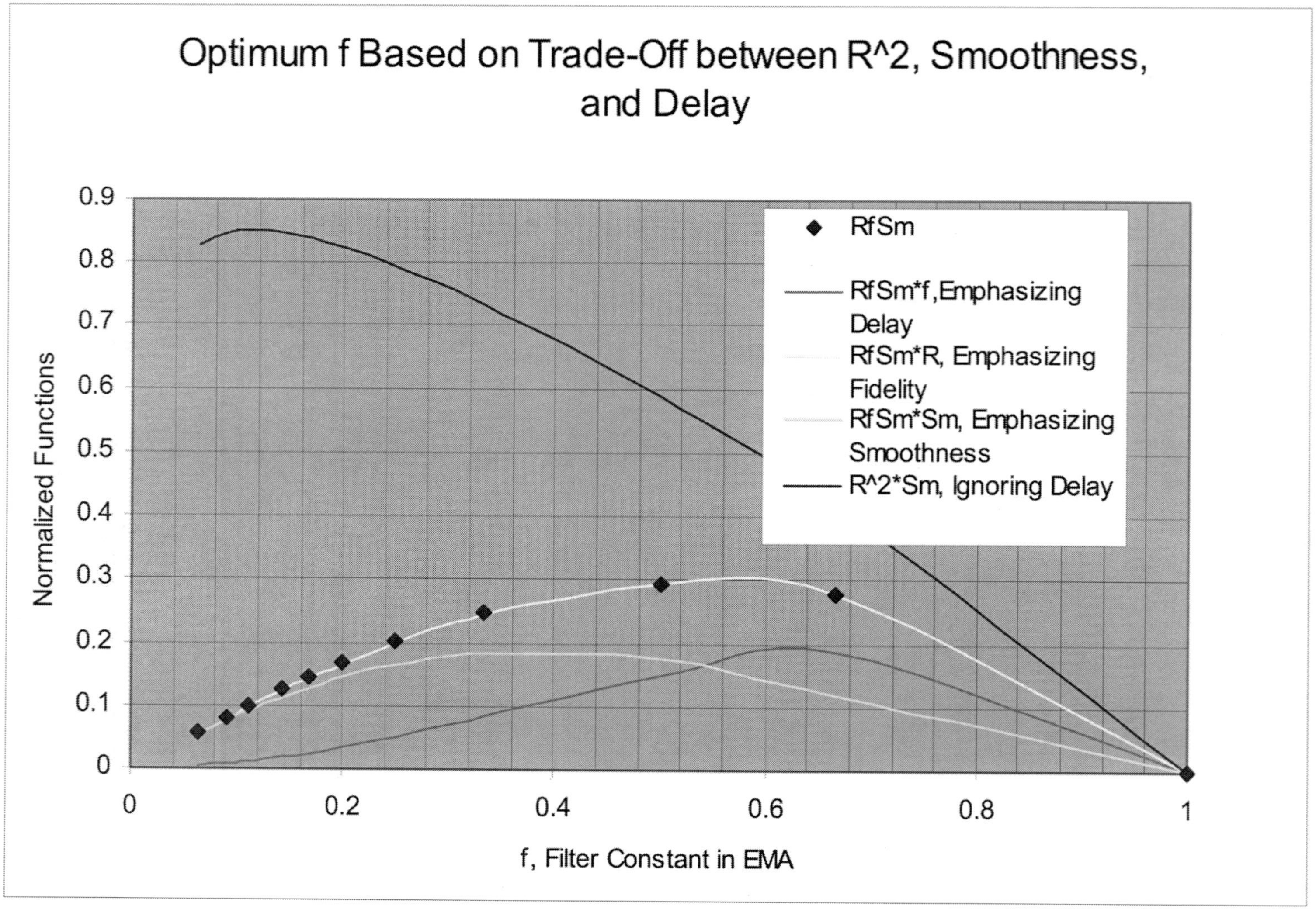

Chen, Lo, Swan, Wetuski, & Stout

Dynamic Regression

- Dynamic regression was utilized to identify parameters in dynamic models.
- Dynamic models
  - autoregressive exogenous input (ARX) model
  - autoregressive moving average exogenous input (ARMAX) model
- Related information such as time constant can be extracted from model coefficients
- The ARMAX model is desirable if unmeasured disturbances exist.

NEXUS ENGINEERING

Chen, Lo, Swan, Wetuski, & Stout

ARX Modeling for Atmospheric Ozone

- Dead times were determined with the Genetic Algorithm (or the cross correlation analysis).
- Time constant was determined with a multiple input single output (MISO) model

NEXUS ENGINEERING

Formation of Ozone

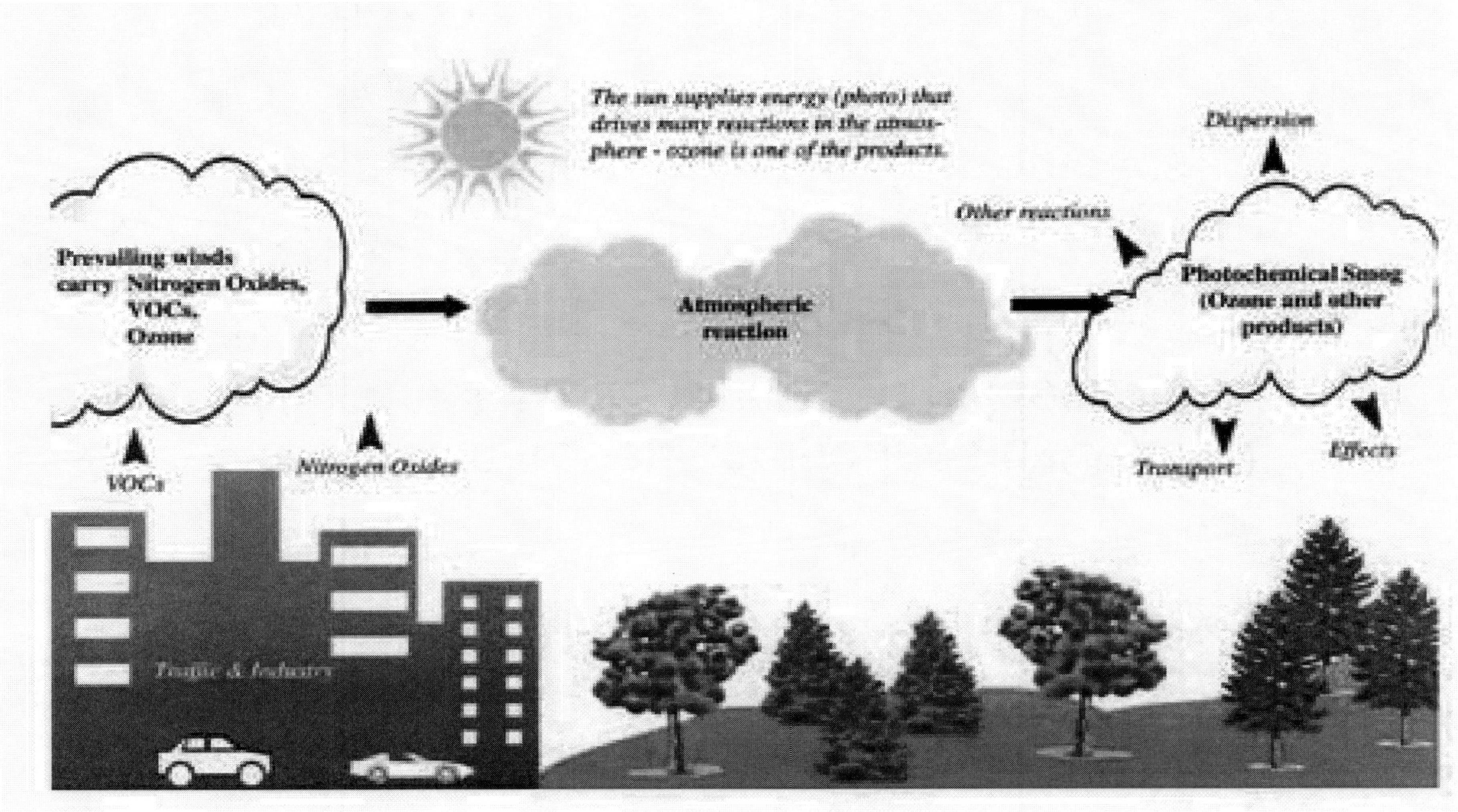

Factors affecting ozone/PM formation

- Precursors
  - Reactive nitrogen compounds (NO_y): NO_x ($NO + NO_2$), other nitrogen oxides(NO_3, N_2O_3, N_2O_5), HONO, HNO_3 , HNO_4
  - SO_2, NH_3
  - VOCs (Alkanes, Alkenes, & Aromatics)
  - O_3 (for PM)
  - PM (for O_3)

NEXUS ENGINEERING

Chen, Lo, Swan, Wetuski, & Stout

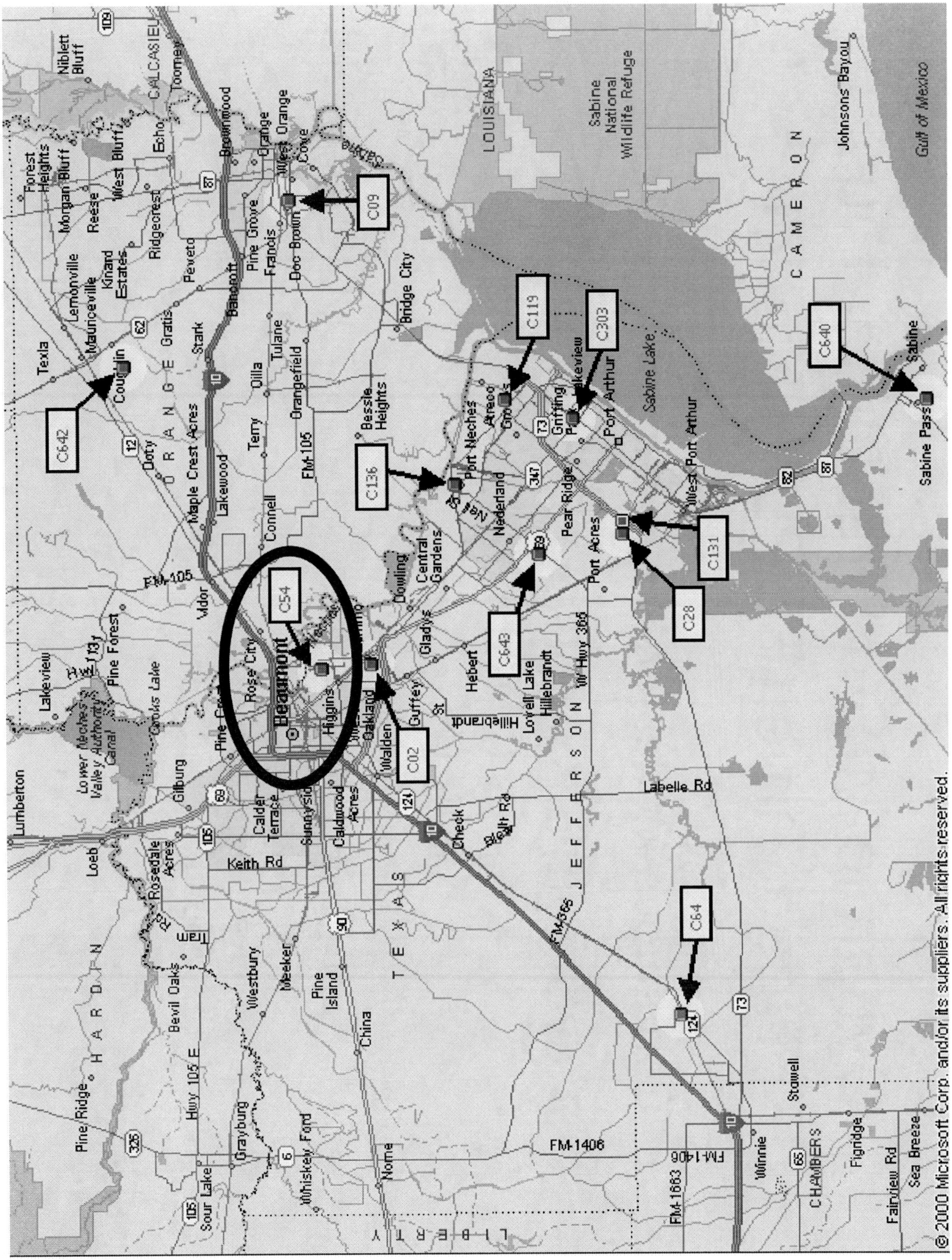
C642
C09
C119
C303
C136
C54
C02
C643
C131
C28
C640
C64
Beaumont
Port Arthur
Sabine Lake
Gulf of Mexico
Sabine National Wildlife Refuge
LOUISIANA
CALCASIEU
C A M E R O N
O R A N G E
J E F F E R S O N
H A R D I N
T E X A S
L I B E R T Y
CHAMBERS
Johnsons Bayou
Sabine Pass
Sabine
West Port Arthur
Port Acres
Pear Ridge
Nederland
Port Neches
Groves
Griffing Park
Lakeview
Bessie Heights
Bridge City
Orangefield
West Orange
Orange
Doc Brown
Vidor
Rose City
Higgins
Oakland
Walden
Guffey
Gladys
Hebert
Lovell Lake
Hillebrandt
Central Gardens
Dowling
Lumberton
Lower Neches Valley Authority Canal
Pine Forest
Lakeview
Rosedale Acres
Calder Terrace
Caldwood Acres
Sunnyside
Keith Rd
Labelle Rd
Cheek
Pine Island
China
Nome
Meeker
Westbury
Bevil Oaks
Pine Ridge
Sour Lake
Grayburg
Whiskey Ford
Winnie
Stowell
Fannett
Figridge
Sea Breeze
Fairview Rd
FM-1406
FM-1663
FM-365
FM-105
W Huy 365
Huy 105 E
Texla
Mauriceville
Lemonville
Kinard Estates
Forest Heights
Morgan Bluff
Niblett Bluff
Reese
West Bluff
Ridgecrest
Peveto
Bancroft
Brownwood
Toomey
Echo
Pine Grove
Francis
Tulane
Terry
Connell
Lakewood
Maple Crest Acres
Doty
Stark
Gratis
Coughlin
Gilburg
Loeb
© 2000 Microsoft Corp. and/or its suppliers. All rights reserved.

Dynamic MISO First Order Plus Dead Time (FOPDT) Model

- Continuous Model for Multiple Inputs

$$\tau\, dy/dt + y(t) = K_{p1}\, u_1\,(t-D_1) + K_{p2}\, u_2\,(t-D_2) + K_{p3}\, u_3\,(t-D_3) + K_{p4}\, u_4\,(t-D_4) + K_{p5}\, u_5\,(t-D_5)$$

where $y(t)$, Measured variable.

u_1-u_5, Output variables NO, T, SR, W and RH respectively.

K_{pi}, Steady state gain.

τ, Overall time constant.

D_i, Apparent Dead time.

Chen, Lo, Swan, Wetuski, & Stout

MISO Discreet FOPDT Model

$$y_k = a_1 \; y_{k-1} + b_1 \; u_{1,k-d1-1} + \ldots + b_m \; u_{m,k-dm-1}$$

where

$u_1, u_2, \ldots \ldots u_m$ = input variables

$a_1 = e^{-\Delta t/\tau}$

$b_1 = K_1(1 - e^{-\Delta t/\tau})$

$b_m = K_m(1 - e^{-\Delta t/\tau})$

τ = time constant

$K_1, K_2, \ldots \ldots K_m$ = gains

$d_1, \ldots \ldots, d_m$ = dead times

Chen, Lo, Swan, Wetuski, & Stout

Significant Variables for Ozone Modeling

- NO
- Temperature
- Relative Humidity
- Solar Radiation
- Resultant Wind Speed

NEXUS ENGINEERING

Chen, Lo, Swan, Wetuski, & Stout

Results of Dynamic Regression

- Our prior works showed high R-squares (0.88 – 0.97 for ozone) with these easy-to-use and computationally efficient models.

NEXUS ENGINEERING

Chen, Lo, Swan, Wetuski, & Stout

Forecast vs observed normalized ozone using Excel hourly dynamic model (3 days)

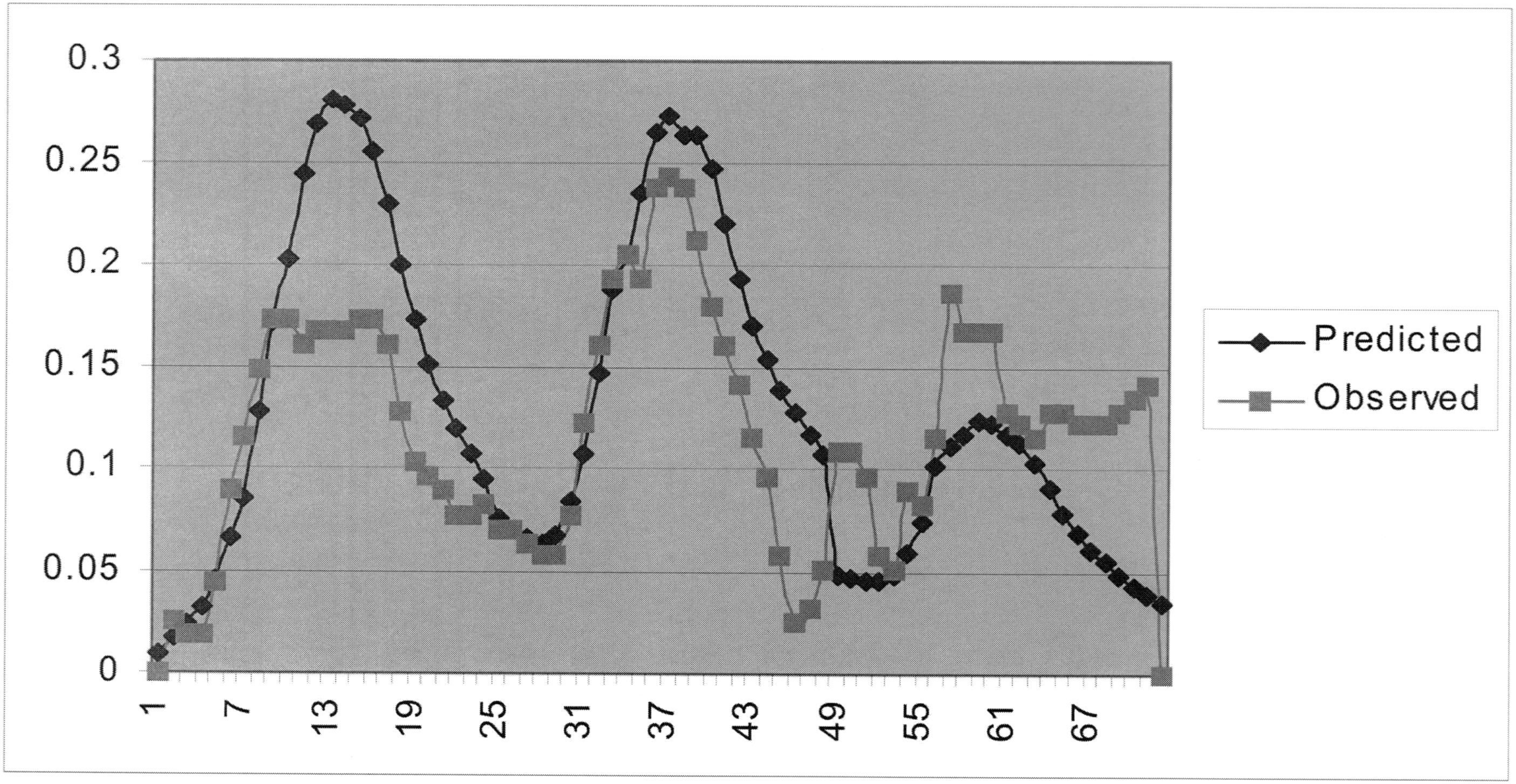

Software

- Lamar University has licenses for several well-established industrial modeling tools such as Matlab, Aspen IQ, Aspen IQ Powertools, Aspen Dynamics, and HYSYS, to support the teaching and research in this area

NEXUS ENGINEERING

Chen, Lo, Swan, Wetuski, & Stout

Thank you!

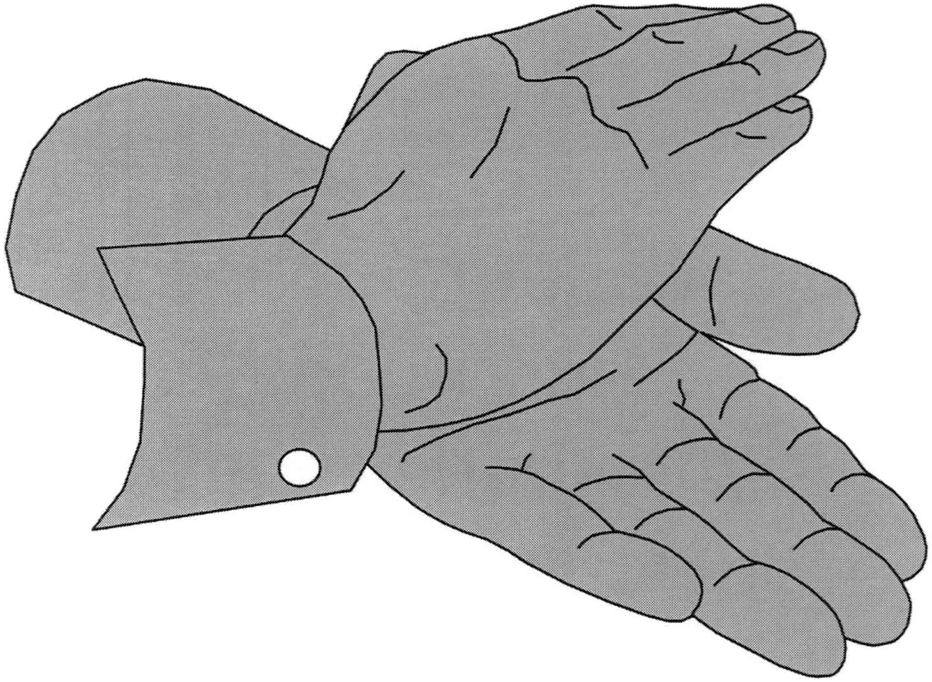

We appreciate your feedback and comments!

Chen, Lo, Swan, Wetuski, & Stout

Information Technology, A Challenge/Opportunity for Chemical Process Industry and Engineering Education

Kuyen Li, Amarnath Singh, and Jack R. Hopper
Chemical Engineering Department, Lamar University, Beaumont, Texas 77710

Abstract

The U.S. chemical process industry (CPI) faces three challenges: global competition, strict environmental and safety regulations, and stringent customer requirements. These challenges create opportunities for the CPI to look into better technology solutions and improve plant operations by using **information technology** (IT). In the CPI, four information management areas are identified: (1) Enterprise Resource Management, (2) Project Management, (3) Process Information Management, and (4) Asset Management. Under current chemical plant operations, operation and maintenance (**O&M**) activities are integrated while health, environment, and safety (**HES**) activities are performed nearly independently from the plant operation. This situation can be improved through a setup of IT infrastructure. Furthermore, the CPI faces scattered sets of OMS (Operation, Maintenance, and EHS) packages from various service providers that are incompatible and hard to integrate.

Lamar University has held an IT seminar series, with emphasis on OMS in the CPI, to bring three distinct groups together: IT solution providers, CPI representatives, and academia. The objective was to see how IT could be used in the CPI to improve its operation, maintenance and safety. The seminar series covered the basic **asset linking technology**, the design and operation of **life cycle** information platform, the OMS fine-tuning for environmental regulation, the Vendor City and **technical library**, and the infrastructure for plant life cycle.

Introduction

The U.S. chemical process industry (CPI) is the world's largest producer of chemicals, contributing the largest trade surplus of any non-defense-related sector to the U.S. economy, representing 10 percent of all U.S. manufacturing, and employing more than 1 million Americans. The CPI faces three challenges: (1) global competition, (2) strict environmental and safety regulations, and (3) stringent customer requirement [1]. The CPI is urged to produce faster, cheaper, and safer. These challenges, however, create opportunities for the CPI to look into better technology solutions and improve plant operations by using information technology (IT). The IT term we are using here refers to the system or ability to move a large amount of data from the field level through out an organization and extract the valuable information in order to maximize their values. The challenge for academic institutions is to understand the complex situation facing the CPI today in order to better educate future CPI leaders. Lamar University is committed to meeting this challenge.

CPI Enterprise Information Environment

The CPI enterprise information environment may be depicted in a pyramid as shown in Figure 1 [3]. The top part of the pyramid is the supply chain management (SCM1) system while the bottom part is what is called service chain management system (SCM2). Any CPI today really operates in a global environment from the supply chain or service chain point of view because these systems connect CPI globally to suppliers, customers, vendors, and regulatory agencies. The SCM1 system in CPI is the enterprise resources management (ERM), which includes accounting and financing management of purchasing, manufacturing, maintenance, and distribution. The SCM2 consists of three sections: (1) project management, (2) process information management, and (3) asset management. The project management section includes engineering, procurement, and construction (EP&C) where much of the plant information data are generated. The process information management section includes process simulation, control (DCS), and optimization where a lot of the manufacturing process data are transferred, stored, and archived. The asset management section, as originally described, includes mainly maintenance and today it also includes environment, health, and safety (EHS).

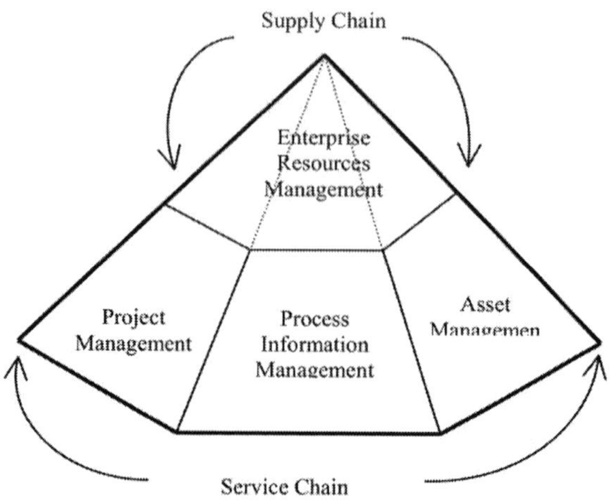

Figure 1. CPI Enterprise Information Environment

The concepts of supply chain management (SCM1) and service chain management (SCM2) have been well established in the CPI. The CPI needs to manage both SCM1 and SCM2 optimally to stay in business. In reality, SCM1 is CPI's core business while SCM2 consists of three sections (see Figure 1), which is supported by service chains such as equipment vendors, contractors, and field service companies. CPI and their service chains are very much interwoven. This complex situation of the SCM2 makes it very difficult to achieve "seamless" service. Because of the extreme complexity of the problem, the ultimate solution will require the utilization of IT to achieve the goal. In the current environment of IT, the infrastructure needed to support common issues under SCM1 and SCM2 is still in a transition stage.

IT in CPI Today

Several solution providers such as ASPEN and OSI Software (PI System) have developed products to promote the vertical connection from the real-time operation data to the supply chain management in order to maximize the profit [4, 5]. Another solution provider, Intergraph, has proposed to provide the engineering design and construction information, and continuously support for plant operation and maintenance [6].

For asset management, the CPI has moved from *corrective* maintenance in the 1970s to *preventive* maintenance in the 1980s and then *predictive* maintenance in the 1990s. The major DCS vendors (such as Honeywell, Foxboro, Emerson, and Yokogawa) are adopting asset management strategies to align with CMMS (computerized maintenance management systems) providers and develop products that optimize data integration. A typical trouble-shooting event begins by lugging test equipment to the devices, which are often located in remote, hard-to-access areas. If spare parts are required, the trip is repeated to conduct the repair. With asset management software, the instrument technician diagnoses the problem from the PC console. One trip completes the repair [7].

For the environmental applications, IT solution providers have concentrated on the application of environmental compliance. PDC (Process Data Control) Corporation has a successful consortium addressing EPA compliance solutions while EMS (Environmental Monitoring Service, Inc.) has successfully demonstrated an implementation of environmental software in the field especially useful in preparing fugitive emission reports. However, not all of the IT solutions have matured at the same rate. Some of the solution providers are not stable. Many of them do not share a common interface, nor do they often connect to the legacy systems.

The CPI is working hard to adapt IT into their operation; however, the culture gap, which exists between the "paper based" and the "IT based" approaches, is still large. Furthermore, the lack of readiness of their personnel and the traditional mind-sets in operations hinder the process and retard the rate of adaptation. As the personnel are accustomed to using outdated, time-consuming procedures to perform various daily functions, these routine procedures are not being converted into an IT environment very rapidly. Moreover, the IT life cycle is usually only 2 or 3 years; which unlike the CPI's life cycle that may span 20 years or more.

Information Technology Seminar Series

The college of engineering initiated an IT seminar series at Lamar University during 2001. We selected OMS (Operation, Maintenance and Safety) as the main theme for the IT seminars because this area is not yet well established, even though the DCS has been used as the backbone in plant operations.

OMS Solution by its very nature provides plant managers with a grasp of both their daily and hourly needs. As one of the four basic areas, OMS imports extensive data/information from EP&C Solution while interfacing with Simulation & Optimization Solution during the life cycle O&M stage. A good communication between OMS and Enterprise Solution makes the CPI global operation on a "consistent basis". So, obviously, OMS Solution enables plant-site operation, maintenance, and HES personnel to enjoy a better communication through a common platform and results in a better return on investment for the CPI. Lamar's Industrial Infrastructure Committee of the Engineering Advisory Council has targeted OMS Solution as the goal of our seminar series. The topics of this seminar series are shown in Table 1.

Table 1. Topics of the Seminar Series "Information Technology, A Challenge in CPI"

1. OMS Basics - Asset Linking Technology (ALT)
2. OMS Within A Big Picture - Design and Operation Life Cycle Information Platform
3. OMS Fine-Tuning with Considerations of EHS (environmental, health, and safety)
4. OMS Extension – Vendor City and Technical Library
5. OMS Infrastructure for Life Cycle
6. Life Cycle Issue with Solution Providers: Software and Hardware

No OMS solution exists today under the IT environment. In fact, under the current U.S. CPI operation, only O&M are operated interactively, while HES groups are working nearly independently. This situation can be improved through the IT infrastructure. However, the CPI faces numerous options of OMS packages from various service providers that are incompatible and hard to integrate. This seminar series offered one step enabling people to look at solutions that are being designed and put into place to facilitate smooth legacy transitions, increase efficiencies and increase profitability [2].

A study of process upset releases has been made by the ERRI (Episodic Release Reduction Initiative), an innovative voluntary effort among the U.S. Environmental Protection Agency (EPA) Region 6, the Louisiana Department of Environmental Quality (LDEQ), the Texas Natural Resource Conservation Commission (TNRCC and now TCEQ, Texas Commission of Environmental Quality), and the regional petroleum/chemical industry to evaluate the causes of releases to the air associated with startups/shutdowns, equipment failures, and process upsets. This study indicated that the major causes were from O&M. In the ERRI final report, it was suggested that operation, maintenance and HES be combined into one system [8]. The report did not provide an infrastructure for combining these three functions.

Experiences from CPI and Its Solution Providers

1. Asset Linking Technology (ALT)

Grant Dawson shared his companies' experiences in the setup and training of an Asset Linking Technology (ALT) at Lyondell-Equistar. With 30 plant sites in North America, Lyondell-Equistar has a very challenging enterprise operating environment. An urgent need is to homogenize and standardize values, business style, culture, infrastructure, and legacy systems. As pointed out by Dan Smith, the CEO of Lyondell-Equistar, "*---It is critical to the success of our organizations that information and training quickly and smoothly flow through the organization. One way we can do this is by very quickly rationalizing inherited IT legacy systems from various components of the company into one working system [2].*"

Assets represent years of accumulated knowledge, practices, and procedures that can be linked together without sacrificing their current roles in plant operations. The key is ALT, a linking engine to connect information from different components. ALT allows plant personnel to access information at their own pace and need, thus creating a true knowledge based worker. ALT also provides plant-site OMS personnel with freedom of collaboration among themselves and other disciplines within plant-sites. This presentation gave a detailed example of computer-based-training (CBT) in the chemical industry [9].

2. Life Cycle Operation

During the second seminar, Bob Humphrey shared Intergraph's vision of "Life Cycle Engineering Information Management". Intergraph is working with proven solution providers to develop a prototype for life cycle operation supported by a data warehouse. This prototype will provide cost-effective solutions for plant life cycle OMS. The presentation pointed out the roots of the data quality problems embedded during EP&C data hand-over and in document-centric data management such as data not linked to asset. The manual MOC (Management of

Change) processes are suitable for the mid-life of O&M but are not for revamps, upgrades, and increased maintenance actions [10].

3. OMS Fine Tuning with Environmental Regulation

The third seminar concentrated on the usage of IT as environmental compliance solutions. PDC (Process Data Control) Corporation has a successful consortium addressing EPA compliance solutions while EMS (Environmental Monitoring Service, Inc.) has demonstrated successfully an implementation of environmental software in the field. This seminar presented how both compliance solutions and the orderly transfer of data/information from plant design become an integral part of plant real-time operation and furthermore benefit the operation and maintenance personnel (O&M and Health-Safety-Environment integration) [11].

This seminar has presented the critical issue of compliance management plan (CMP), which breaks down the OMS compliance responsibilities into "tasks" and "procedures". Define tasks and procedures facilities tracking of compliance and determination of compliance deviation. One advantage of a CMP is that it can be automated to increase efficiency. The seminar also gave examples of CMP at task levels for air compliance.

4. Vendor City

PPG shared their 14 years of experience in Vendor City operation in Lake Charles, Louisiana. This presentation provides a good example of how future OMS can be chartered with service chain management (SCM2) providers. SCM1 (Supply Chain Management) and SCM2 are both needed for the CPI. Their interwoven nature is illustrated in Figure 1. As SCM2 involves many different types of business, this seminar has presented a focus on vendors' role. IT enables a better management of SCM2. For an example, the mean time before failure (MTF) is what vendors claim they can offer, but what vendors really offer after sales is mean time before fixing (MTF2). These scorecards need CPI's support together with a usage of information technology in the CPI [12]. A third party, other than CPI and vendors, could be a good place to keep and manage these scorecards.

5. From NASA Sub-contractor to CPI Solution Providers

Dealing with the rapid evolution of Information Technology, both in software and hardware, has been a nightmare for the CPI information system (IS) departments. Moreover, compatibility issues hinder the progress of IT integration. This common problem has not been resolved yet in the CPI. NASA had a similar problem and has relied on C3I (Command, Control, Communications, and Intelligence) provided by LinCom to move from the 60's to current operation with continuity in IS issues.

LinCom's experiences may provide the CPI with a good roadmap to deal with this IT infrastructure issue. LinCom also would like to transfer their well-developed information technology into the CPI. The transfer may not be easy because of the different cultures/environments between NASA and CPI. Lamar may be able to serve as an incubator for this technology transfer [13].

6. <u>Conclusion from the Seminar Series</u>

Many plants have a life span of over 30 years, while the life cycles for IT software and hardware are only a few years. This can be a problem for the adaptation and implementation of IT in the chemical industry. The sixth seminar concluded the IT Seminar Series at Lamar and summarized the current solution providers' maintenance and upgrading practices in software and hardware. There are four basic Solutions provided by supply and service chains firms which support the CPI plant information Life Cycle need: (1) Enterprise Solution (such as SAP), (2) EP&C Solution (such as Intergraph), (3) Plant Operation (optimization and simulation) Solution (such as AspenTech), and (4) OMS Solution (no leading package).

Enterprise asset management incorporated with EHS functions has been the "gray area" for the CPI. Traditionally, OMS (operation, maintenance, and HES) personnel are not well integrated. There are also several service providers, which supply legacy systems that are not well interfaced. Under a normal condition, the OMS Solution (IT tool) would import most of the plant data/information from the E&C Solution. During the plant life cycle, the OMS Solution is frequently updated through Plant Operation Solution, and would report constantly to and receive command from Enterprise Solution. Furthermore, the OMS Solution, which links to current legacy systems, can provide OMS personnel quick access to the data/information they need to operate the plant in an optimal as well as a compliant status in real time environments. It is time for the CPI to recognize the importance of working together with the Supply and Service Chains in a joint effort to overcome the challenge of IT. Today, it is not enough to access a vast quantity of information; but something productive must result from this knowledge and there is no better time for this to begin than the present [14].

IT Issues and Concerns in CPI

1. <u>Modern Control Architecture in CPI</u>

With the OPC (OLE for Process Control), which was created by Microsoft for process control, the field data will be able to communicate with all kinds of devices and control systems in the chemical industry [15, 16]. This data communication can be accelerated by the use of the Fieldbus in modern process control and asset management architecture in the chemical industry. Much of the functionality contained in DCS, such as deterministic scheduling, a distributed database, and remote alarming, now can migrate to field devices and the network itself. Many key HMI (Human Machine Interface) functions, such as operator interface, system configuration, maintenance, and diagnostics, can be performed by low cost PCs. The gap between product quality, production cost, scheduling to operation and the process variables could be minimized or even eliminated. People may doubt the usage of the Fieldbus in the future of the chemical industry. It probably will not happen quickly but there should be no question that the Fieldbus will be more popular in the future [17].

2. <u>Data Screening and Analysis</u>

With a large quantity of data moving across the field at a high speed, it is very easy for the system to become buried and bogged down. Therefore, data screening and analysis will be very critical; not only for the production management but also for the plant engineer. Before data is transferred to the next level, it is always wise to screen the data to make sure that it is reliable and the quantity is reduced. Statistical data analysis and pattern recognition could be quite helpful for screening the data. Regression and Neuro-Network analysis may be used to find a correlation between product quality and process variables. This correlation will be very beneficial for production management in the chemical industry.

3. Culture Gap and Human Factor

Dealing with the rapid evolution of information technology, both in software and hardware, has been extremely frustrating for the CPI information system (IS) departments. These frustrations could be rooted to a culture gap between IT and CPI industries. A renewal or upgrade of IT software and hardware are quite common in computer business while for the CPI, this is a nightmare. Quite often, a CPI plant manager with one bad IT experience will hesitate to accept the usage of IT any further. This causes a hindrance in the progress of IT usage in CPI.

The human factor is certainly an important factor in the development and utilization of information technology. It must be considered at the very beginning of any culture transition. Mike Aldredge, former Chairman of Puffer Sweiven, addressed this issue in his remarks to the seminar: "*---There is no doubt that in time Information Technology can and will positively impact the OMS portion of Service Chain Management. Sure there are some technical and compatibility issues but those will be solved. My greatest concern is not a technical issue but rather a human factor issue. It is my belief that precious time will be lost if we do not address the human factor issues regarding implementation at the same time that we address the technical issues.*

Integrated Asset Management systems have been around for a number of years with varying degrees of acceptance and success. Why does an O&M package that works great at a plant in Texas fail miserably in an identical process unit in England? What is the reason that an O&M package, which works great in Germany, performs only marginally satisfactorily on an identical plant in the U.S.?

I propose to you that there are un-addressed "turf" issues and these issues will be stumbling blocks until remedied. The "turf" issues may be between individuals, departments, or companies. It makes little difference where the issue is located; the fact is the "shoe is in the gears" and so long as we allow this situation to exist we are sabotaging opportunity.

We must learn to tear down both inter and intra company fences that have been built around the "sacred turf" and encourage all parties to build the bridges necessary for smooth integration. Then and only then, once the technical issues are behind us, can implementation occur unimpeded and the anticipated benefits be derived."

Conclusion

Four information technology areas have been identified in CPI: Enterprise Resource Management, Project Management, Process Simulation and Optimization, and Operation, Maintenance and Safety (OMS). The implementation of IT in all these four areas is quite popular except the OMS area. It is clear that information technology is a powerful tool to enhance the operation and management of CPI to meet the challenges today and tomorrow. Some technical problems will be solved sooner or later. However, the issues of culture gap and human factor remain with little or no attention.

References

[1] Technology Vision 2020, The U.S. Chemical Industry, a report issued by the American Chemical Society, the American Institute of Chemical Engineers, the Chemical Manufacturers Association, the Council for Chemical Research, and the Synthetic Organic Chemical Manufacturers Association, and available online at *http://www.oit.doe.gov/chemicals/page10.shtml, December 1996.*

[2] K. Li, J. Gossage, H. Ortega and H. Lee, "Information Technology, A Challenge to Chemical Process Industry and Lamar University," in the Proceeding of the 2nd International Conference on Information Technology Based Higher Education and Training, pp. 317-321, July 4-6, 2001, Kumamoto, Japan.

[3] S.H. Lee, private communication, 2001

[4] L. B. Evans, "The Digitized Plant," Chemical Engineering, March 2002, pp. 43-46.

[5] *www.osisoft.com/PIsystem*

[6] www.Intergraph.com/pbs

[7] J. S. Gerold, "No-Sweat Asset Management," Control Engineering June, 1997

[8] ERRI Final Report (*www.utexas.edu/research/ozonespikes/episodicEmiss. RedRpt.pdf*)

[9] Grant Dawson, Seminar No. 1, OMS Basics - Asset Linking Technology (ALT) using MocKingbird® Tool, *http://hal.lamar.edu/%7Eche_dept/*

[10] Robert Humphrey, Seminar No. 2, OMS Within A Big Picture - Design and Operation Life-Cycle Information Platform, *http://hal.lamar.edu/%7Eche_dept/*

[11] Jerry O'Brien and Brent Pruet, Seminar No. 3 OMS Fine-Tuning, *http://hal.lamar.edu/%7Eche_dept/*

[12] Lewis Pratt, Seminar No. 4, OMS Extension - Vendor City and Technical Library, *http://hal.lamar.edu/%7Eche_dept/*

[13] Bob Culpepper and Bob Phillips, Seminar No. 5, OMS Infrastructure for Life Cycle, *http://hal.lamar.edu/%7Eche_dept/*

[14] A. Caceres, Seminar No. 6, Life Cycle Issue with Solution Providers: Software and Hardware, *http://hal.lamar.edu/%7Eche_dept/*

[15] M. M. Mansy, G. K. McMillan and M. S. Sowell, "Step Into the Virtual Plant," CEP February 2002, pp. 56-61.

[16] B. L. Braunschweig, C. C. Pantelides, H. I. Britt, and S Sama, "Process Modeling: The Promise of Open Software Architectures," CEP, September 2000, pp.65-76.

[17] D. A. Glanzer, "Take Process Control to a New Level," CEP, October 1998, pp. 23-30.

IBM Software Group

From Automation, Integration to Collaboration: The Business Values, Issues and Approaches

Son X. Huynh
Solutions Executive
Chemical & Petroleum Industries
Industrial Sector
huynhson@us.ibm.com

AIChE
Houston, April 26-31, 2004

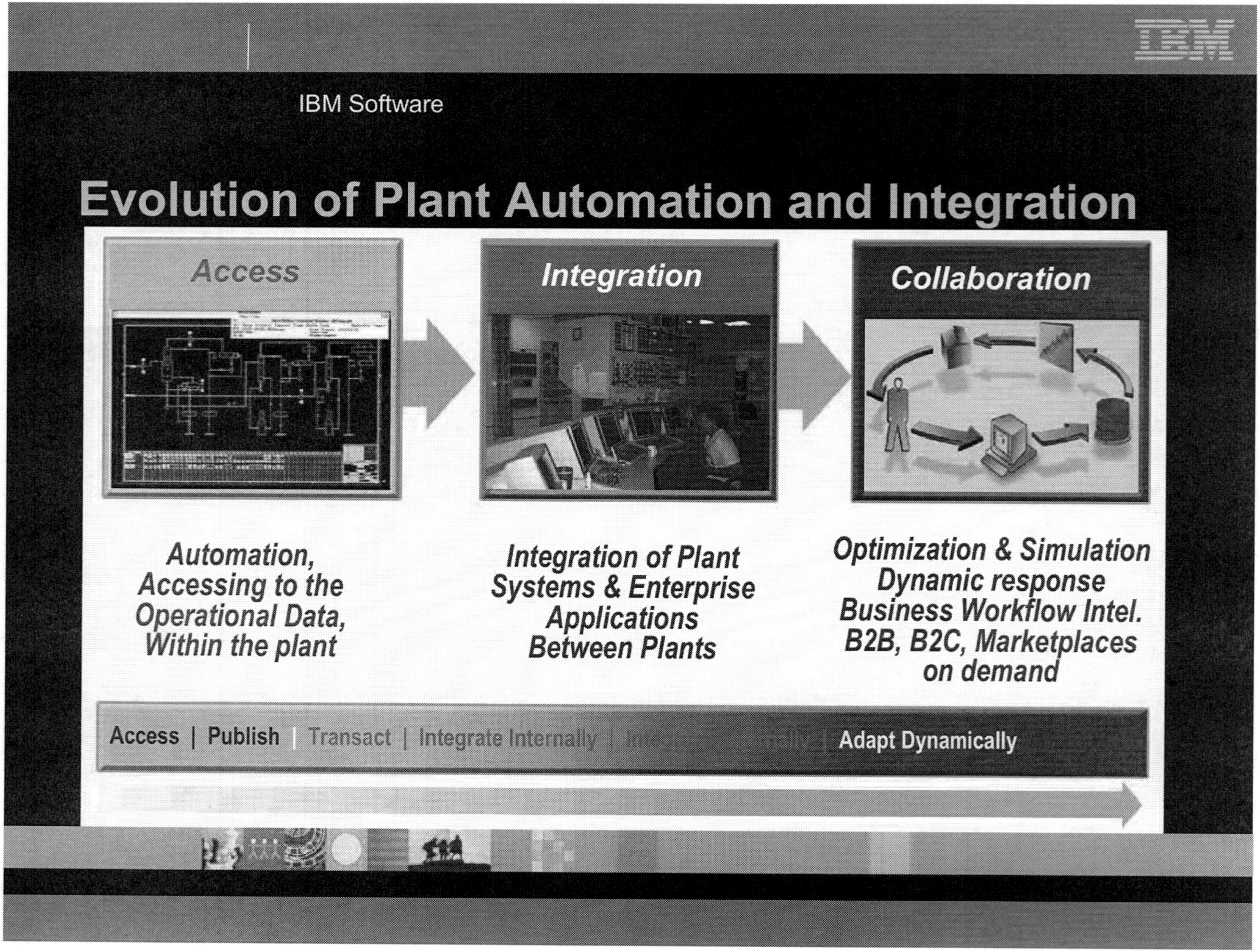
IBM Software
Evolution of Plant Automation and Integration
Access
Integration
Collaboration
Automation, Accessing to the Operational Data, Within the plant
Integration of Plant Systems & Enterprise Applications Between Plants
Optimization & Simulation
Dynamic response
Business Workflow Intel.
B2B, B2C, Marketplaces on demand
Access | Publish | Transact | Integrate Internally | Adapt Dynamically

IBM
IBM Software
The Gap Theory
Other Business Legacy Systems
e-Business / B2B & MarketPlaces
CRM
Business
BW
ERP
APS/SCM
The Boundaries of Information, Culture, Technology and Focus
FIELD DEVICES
Smart Devices, Field Bus
QUALITY
LIMS
Lab Instrument, Test Devices
PROCESS
PIMS
DCS, PLC, NC,..
MATERIAL
Data Collection
Barcode, Scales, RF Devices
PLANT LEGACY SYSTEMS
Engineering & Reporting Systems
Plants

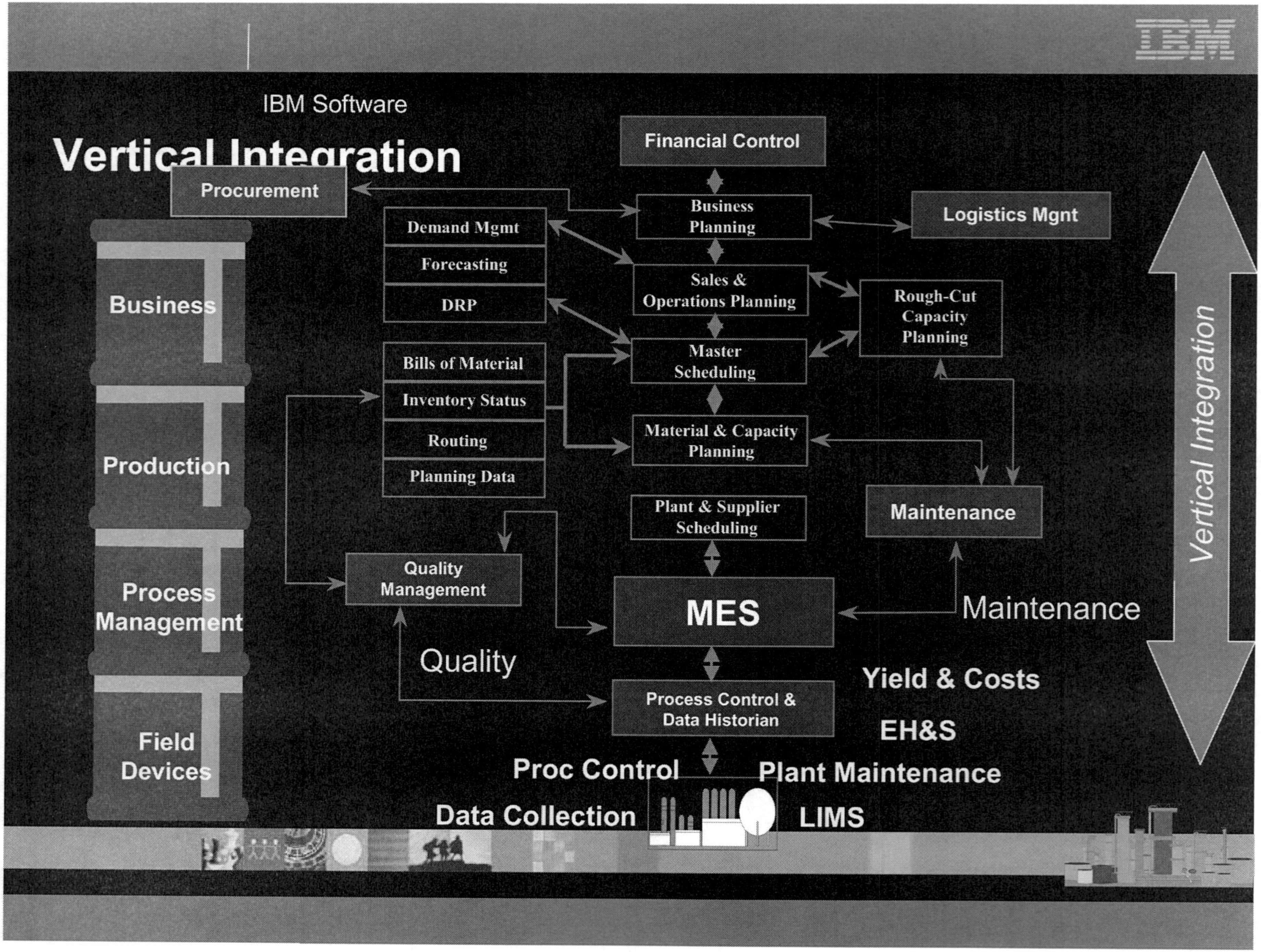
IBM Software
Vertical Integration
Business
Production
Process Management
Field Devices
Procurement
Financial Control
Business Planning
Logistics Mgnt
Demand Mgmt
Forecasting
DRP
Sales & Operations Planning
Rough-Cut Capacity Planning
Master Scheduling
Bills of Material
Inventory Status
Routing
Planning Data
Material & Capacity Planning
Plant & Supplier Scheduling
Maintenance
Quality Management
MES
Maintenance
Quality
Process Control & Data Historian
Yield & Costs
EH&S
Proc Control
Plant Maintenance
Data Collection
LIMS
Vertical Integration

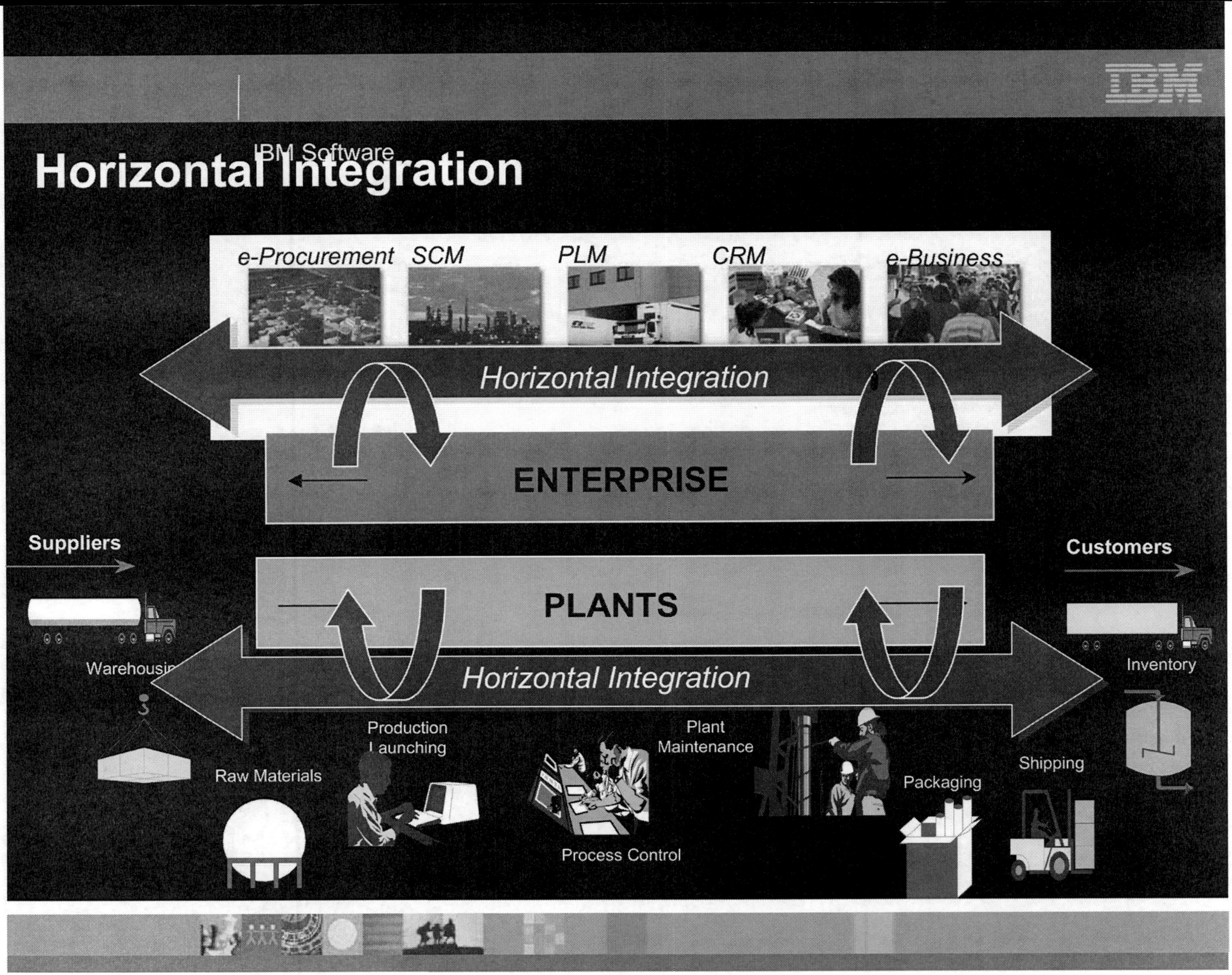
IBM Software
Horizontal Integration
e-Procurement
SCM
PLM
CRM
e-Business
Horizontal Integration
ENTERPRISE
Suppliers
Customers
PLANTS
Horizontal Integration
Warehousin
Inventory
Production
Launching
Plant
Maintenance
Raw Materials
Shipping
Packaging
Process Control

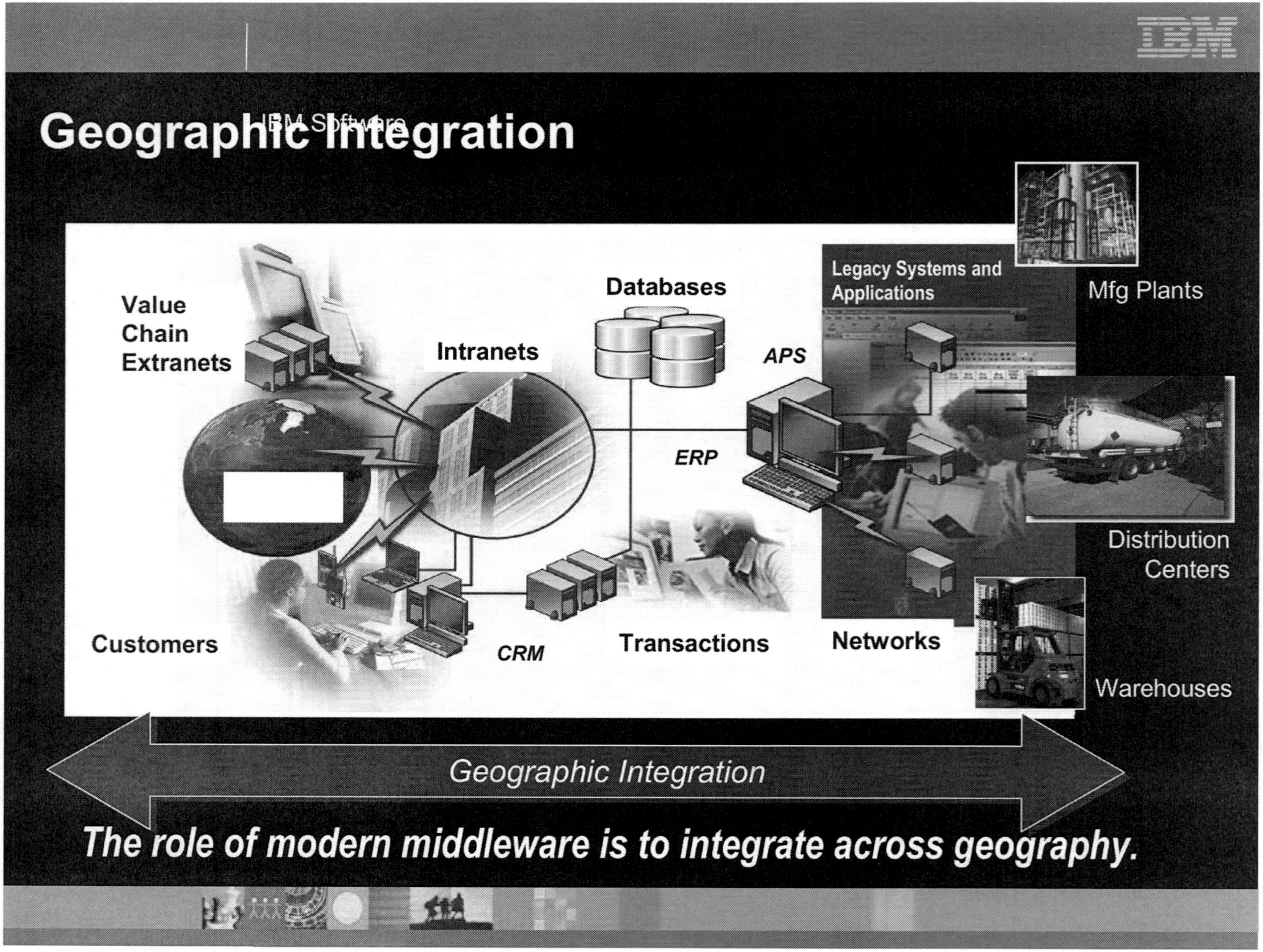
IBM
IBM Software
Geographic Integration
Value Chain Extranets
Intranets
Databases
APS
ERP
Legacy Systems and Applications
Mfg Plants
Distribution Centers
Warehouses
Customers
CRM
Transactions
Networks
Geographic Integration
The role of modern middleware is to integrate across geography.

IBM Software

Evolution of Integration to Collaboration Architecture (SOA)

POINT TO POINT → HUB AND SPOKE → MESSAGING BUS → SERVICE-ORIENTED ARCHITECTURE

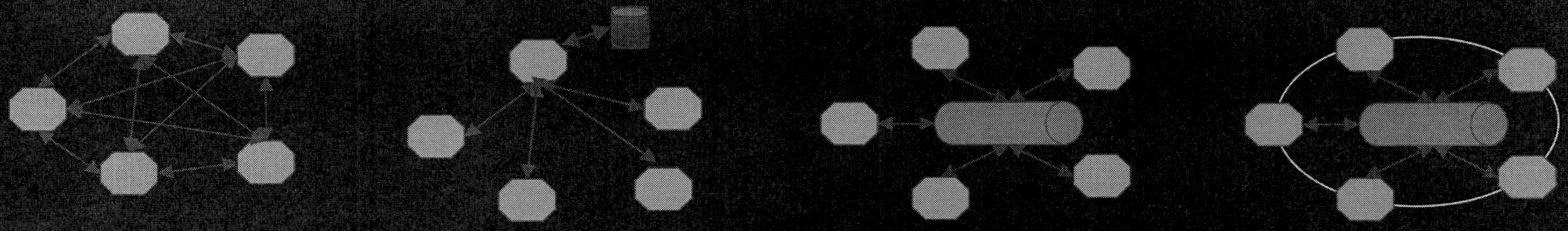

POINT TO POINT

- Existing technologies
- Complex: N*(N-1) Interface Components
- High Maintenance & Support
- Data Passing Not Integrated
- Significant Limitation for Enterprise Integration

HUB AND SPOKE

- Easy to expand
- Required data redundancy
- Data away from the generating sources
- Synchronized based on data but not applications
- High database maintenance

MESSAGING BUS

- Any-To-Any Connectivity
- Publish & Subscribe Tech.
- Distributed & Scalable
- Rich Delivery Functionality
- Offer Multi-Plant/Enterprise Integration
- Well positioned e-Business foundation
- Generic messages are not well defined yet
- Lack of Workflow Mgt and Content Management

SERVICE-ORIENTED ARCHITECTURE

- Tight Integration with BPM
- Enhanced Security//Control
- Persistent Presence
- Push not Pull Technology
- Dynamic Content
- Information Clean-up
- Business Intelligence Workflow
- Alerts and Notification
- Mobile Workforce
- Simulation and Optimization

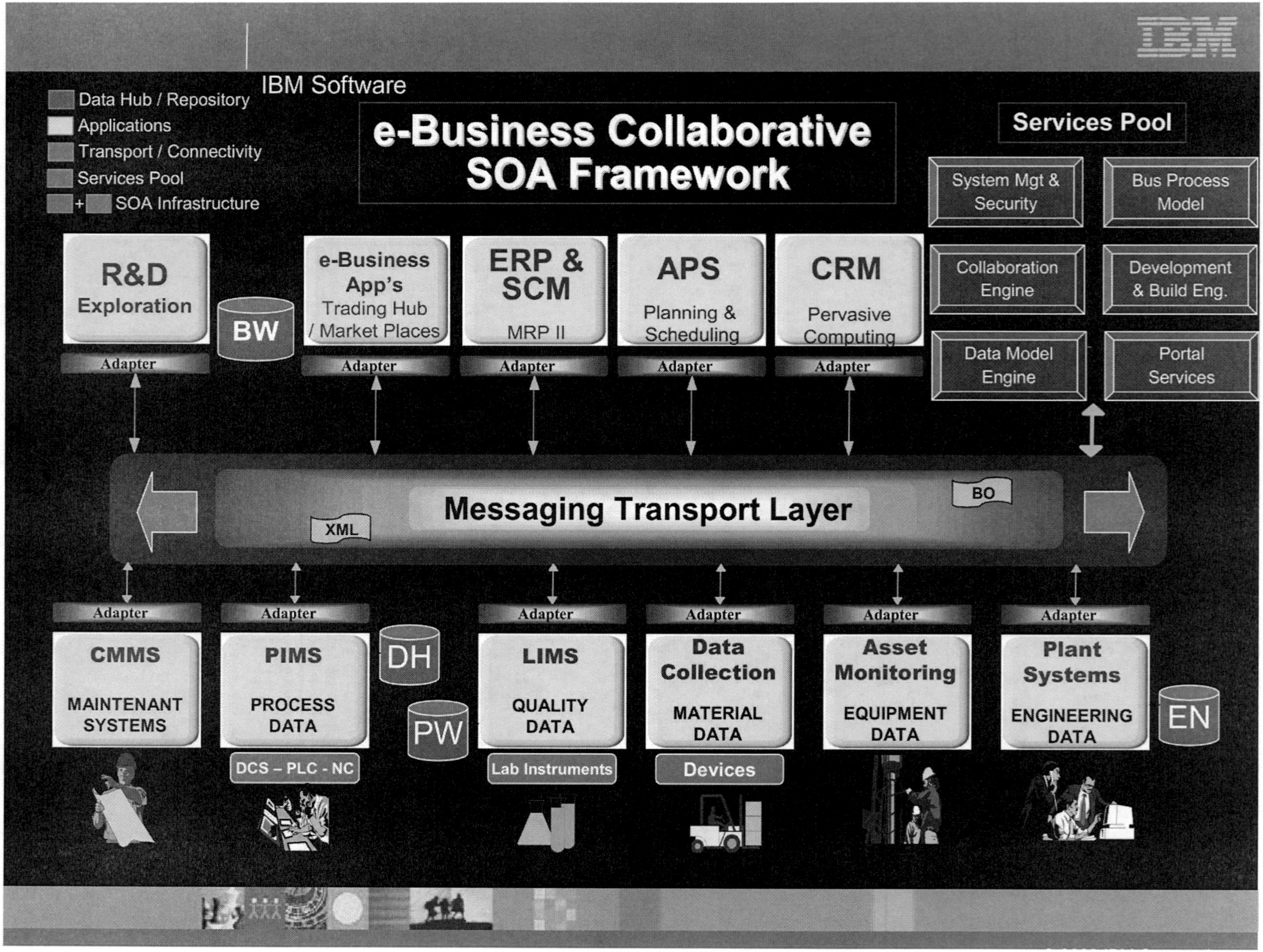
IBM Software
Data Hub / Repository
Applications
Transport / Connectivity
Services Pool
+ SOA Infrastructure
e-Business Collaborative SOA Framework
Services Pool
System Mgt & Security
Bus Process Model
Collaboration Engine
Development & Build Eng.
Data Model Engine
Portal Services
R&D Exploration
BW
e-Business App's Trading Hub / Market Places
ERP & SCM MRP II
APS Planning & Scheduling
CRM Pervasive Computing
Adapter
Messaging Transport Layer
XML
BO
CMMS MAINTENANT SYSTEMS
PIMS PROCESS DATA
DCS – PLC - NC
DH
PW
LIMS QUALITY DATA
Lab Instruments
Data Collection MATERIAL DATA
Devices
Asset Monitoring EQUIPMENT DATA
Plant Systems ENGINEERING DATA
EN

PLANTS mean ..

Why Plants are different than Corporate IT ?

- 7x24 operations environment
- More complex & distributed - Heterogeneous OS, Networks, Applications or Versions - Closer to physical equipments – Many “legacy”, “custom”, “ad-hoc”, “one-of” app’s.
- Designed for real time, event-based, efficiency & time-critical response.
- Systems Security sometimes conflicts in the design, operation and support of the plant manufacturing & control systems.
- High volume traffic, small data package, incomplete and un-constructed format.
- Required specialized and limited resources

Plant IT vs. Corporate IT

IBM Software
IBM
Plant CyberSecurity is about
managing risk, not eliminating it
Risk = Vulnerability x Threat x Consequence
David Kepler – VP & CIO at DOW
"Security considerations extend beyond threat of cyber-terrorism and business continuity to managing outcomes (of): Reliability, Operational performance and Financial performance."
ISA 2003 EXPO – Houston, TX

IBM Software

Hard Facts About Security

- Global cost of cybersecurity breaches is estimated at $1.5 trillion (*Information Week* 2001). Last year, US Corporations lost $59B in proprietary information and IP

- 90% experienced security lapse in past 12 months. Security incidents have quadrupled in two years

- The President's National Strategy for Cyber Security, the chemical and petroleum industry sector, has mandated that I/T Security (Cyber) be added to the Responsible Care Program guidelines for self regulation.

- A study by the FBI and the Computer Security Institute on Cybercrime, released in 2000 found that 71% of security breaches were carried out by insiders.

Security is rapidly advancing as a business priority… in Manufacturing

Strategic Initiative	Scores
Security	5.57
Web Services	4.91
Data Warehousing	4.54
IT Arch Re-Design	4.51

Source: Gartner Server of Key Manufacturing IT Decision Makers, January 2003

7 = high
1 = low

How important is plant security to your company?

100 % Said Plant Security Was Important or Critical

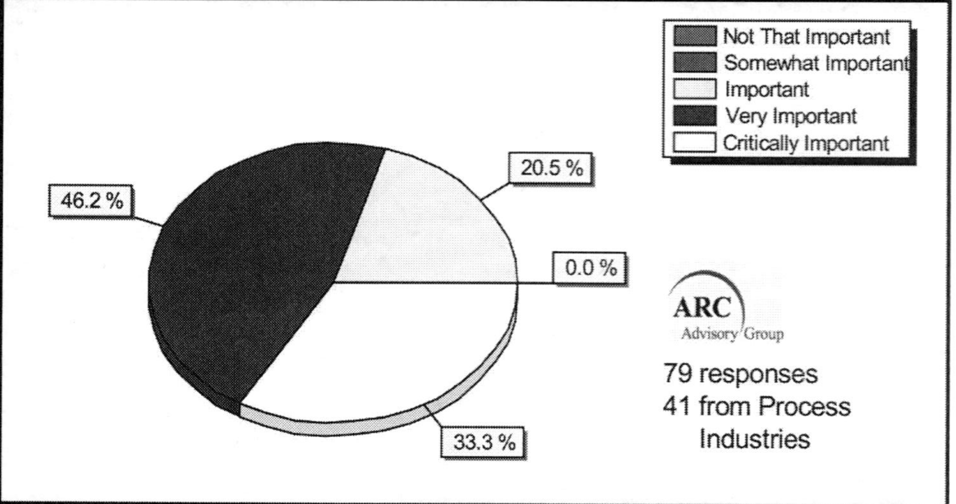

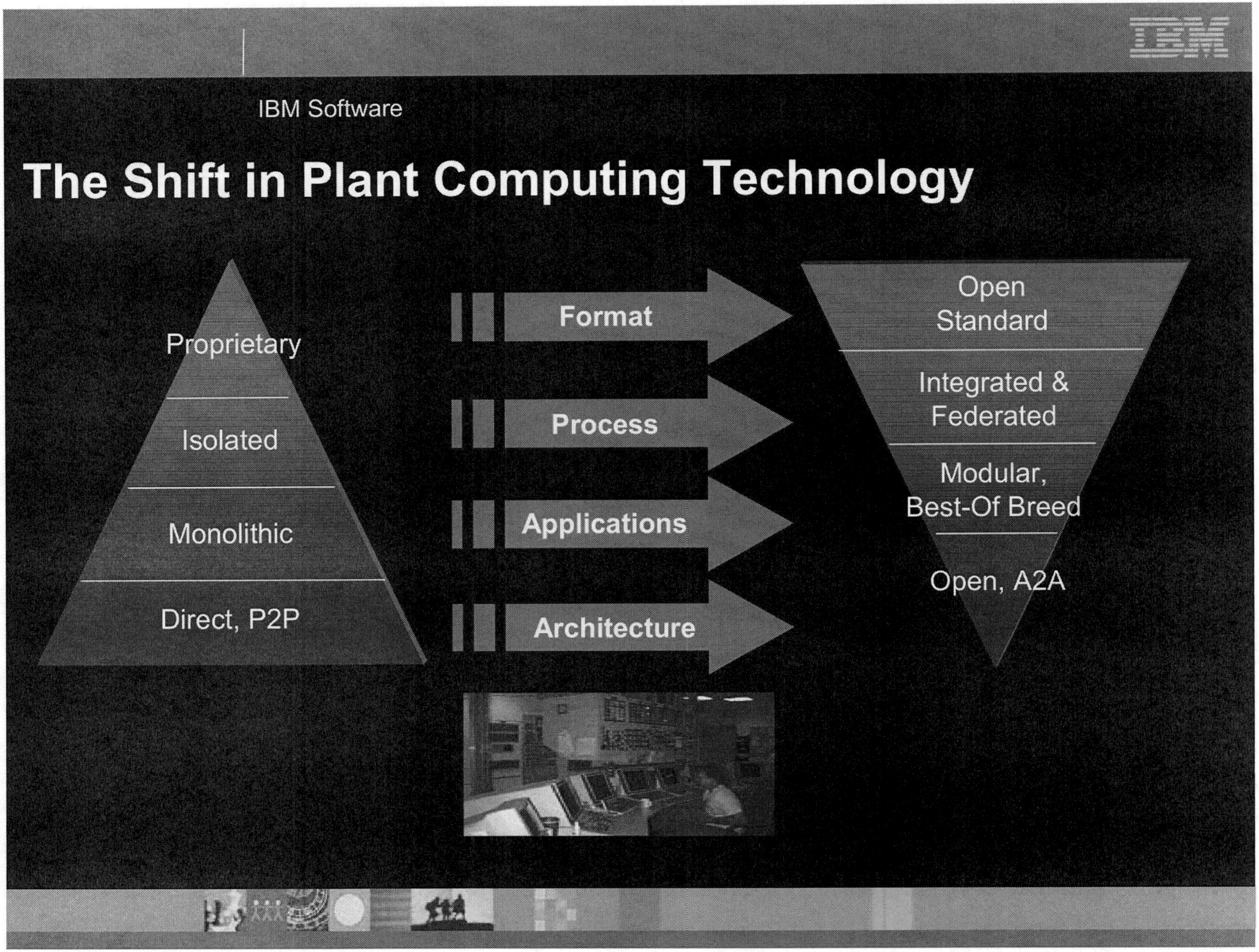
IBM Software
The Shift in Plant Computing Technology
Proprietary
Isolated
Monolithic
Direct, P2P
Format
Process
Applications
Architecture
Open
Standard
Integrated &
Federated
Modular,
Best-Of Breed
Open, A2A

Process Plant CyberSecurity Components

IBM Software

- **Production Operations include**: Conveyance, Treatment, Processing, Manufacturing, Control, Maintenance, Quality Assurance, Distribution and Plant Engineering.
- **CyberSecurity deals with**: Identification, Authentication, Accountability, Authorization, Auditing and Tracing, Administration and Physical (Systems) Security
- **Plant Manufacturing Systems cover**:
  - Personnel, Procedures, Hardware, Software (OS, Applications and Middleware).
  - Standalone, networked or remote access / mobile computing

IBM Software

The Changing Roles of Process & Plant Engineers

'80s

- PID Loops
- Advanced Proc Control
- Plant Design
- Start-Up & Turn-Around
- Project Engineering
- PLC, DCS, Data Historian
- Analyzers, Field Devices
- Process Optimization
- Process Troubleshooting
- SOP's
- ….

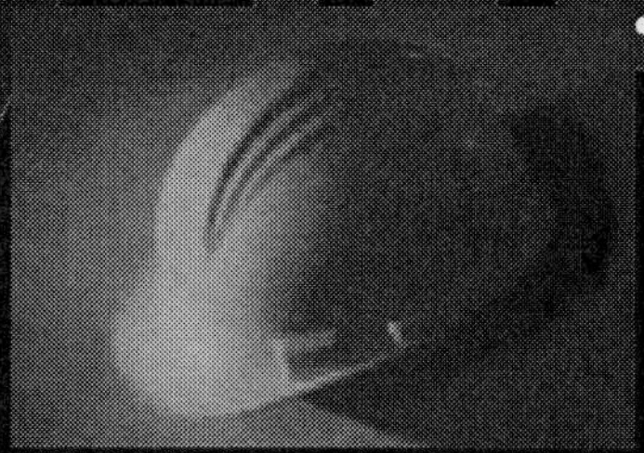

'00s

- **System Integration**
- Six-Sigma Initiatives
- **Supply-Chain Event Management**
- **Plant Cyber-Security**
- Security Patch Management
- Document Management
- Quality / Regulatory Compliance
- Safety
- **Product Costing**
- Customer Technical Support
- **Manufacturing Execution**
- Vendor Profiling
- **ROI Delivery & Mfg Excellence**
- ….

The End

Achieving Lowest-cost Emission Level Compliance Under Cap & Trade Using a Math Programming-based Emissions Optimizer that Links Regional Customers through a Web-based Environmental Management Information System

Derrick P. Schertz and Joseph F. Pekny
Advanced Process Combinatorics, Inc.
West Lafayette, IN 47996

*Richard E. Mc Allister, Beth Brandes Jenkins, Bruce E. Snyder, Semra Alici, Edward J. Galligan, Carlos A. Valenzuela, and Mark M. Daichendt**
Air Products & Chemicals, Inc.
Allentown, PA 18195

ABSTRACT—The Houston-Galveston Area (HGA) is classified as a "serious" ozone non-attainment area. Each company operating in the area must achieve specific NO_x emission level limitations, staged between 2003 and 2008. Various treatment technologies are available for the different types and sizes of emission point sources. Air Products & Chemicals, Inc. (APCI) has developed a unique Services & Solutions Business offering, built on a patent-pending business method, which establishes a web-based emissions network that links companies' emission-reduction plans to determine the lowest cost of compliance under the Mass Emission Cap & Trade (MECT) program. The web-based management system integrates data from the Texas Commission on Environmental Quality (TCEQ) and the customers' forecasted point source emission levels to provide input to an Emissions Optimizer (EO). The EO develops a solution using state-of-the-art mathematical-programming-based optimization techniques. A large-scale Mixed-Integer Linear Programming (MILP) problem is solved using the *NO_x Model Manager* application, customized for APCI by Advanced Process Combinatorics (APC) using APC's proprietary *VirtECS* solver technology, to determine the lowest compliance cost for the region. A subsequent Nonlinear Programming (NLP) problem is solved using the *NO_x Trading Model*, developed by APCI using the *Generalized Algebraic Modeling System* (*GAMS*), to determine the corresponding credit trades. Optimization results are returned to the database for each customer's individual review.

KEYWORDS—Emissions Optimizer, Cap & Trade, Environmental Management Information System, NOx Emission Compliance Monitoring, Houston-Galveston Area, Texas Commission on Environmental Quality, Enterprise-wide Optimization, Regional Optimization, Mixed-Integer Linear Programming, Nonlinear Programming

Introduction

The Houston-Galveston Area (HGA) is classified as a "serious" ozone non-attainment area. Over 2750 point sources are subject to treatment to reduce NO_x emission levels (ton/year) by 80%. Based on current point source activity levels (MMBTU/hr) and emission factors (lb NOx/MMBTU), each company must achieve specific NO_x emission level limitations ("allowances"), staged between 2003 and 2008. Various treatment technologies are available for the different types and sizes of units. These technologies have different emission-reduction

---

*To whom all correspondence should be addressed.

capabilities (lb NO_x/MMBTU) and investment costs ($-hr/MMBTU). In many cases, two technologies can be combined.

Air Products & Chemicals, Inc. (APCI) has developed a unique Services & Solutions Business offering, built on a patent-pending business method, which establishes a web-based emissions network that links companies' emission-reduction plans. APCI entered into an alliance agreement with VisionMonitor Software LLC, who used their Compliance Intelligence suite of products to develop the Environmental Compliance Competitive Advantage (ECCA) solution. The ECCA solution enables organizations to track, monitor, and predict their compliance with various regulatory standards and reporting requirements across the enterprise.

Additionally, the ECCA solution integrates data from the Texas Commission on Environmental Quality (TCEQ) and the customers' forecasted point source emission levels to provide input to an Emissions Optimizer (EO), as highlighted in D'Aquino (2003). The EO determines the lowest cost of compliance with the Texas State Implementation Plan under the Mass Emission Cap & Trade (MECT) program, as described by Huston *et al.* (2002). The EO develops a solution using state-of-the-art mathematical programming-based optimization techniques.

The EO consists of two components: (1) a large-scale, Mixed-Integer Linear Programming (MILP) problem to determine the lowest cost of compliance, and (2) a large-scale, Nonlinear Programming (NLP) problem, to determine the trades between companies corresponding to the investment selections determined by the MILP. These two components will be discussed in the next section.

Air Products has applied for a patent to protect this business model. The model is configured to handle any type of emission data in any non-attainment geography.

Air Products' Emission Optimizer

Mixed-Integer Linear Programming Component

The selection of the lowest cost of compliance for the entire HGA involves discrete decisions. These are, for each point source, when to install what remediation technology or combination of technologies over a six-year time horizon. These discrete decisions are represented by integer (0/1) variables. The lowest-cost solution to this question can be found by formulating an MILP problem. A high-level abstraction of the MILP problem is:

Minimize *Total Discounted Investment Cost*

Subject to:
Emission Level Constraints
Customers' Point Source Emission Forecasts
Technology Emission Level Reduction Capabilities
Customer Fixed Technology Decisions

Large-scale MILPs are considered intractable, because the number of potential solutions to an MILP problem grows exponentially with the number of integer variables. While there is no known algorithm that can solve general MILP problems in polynomial time, previous projects with Advanced Process Combinatorics, Inc. (APC), a venture spin-off company from Purdue University, had shown that, through careful problem formulation, decomposition techniques could be applied to exploit problem structure, and algorithms could be engineered to solve the specific class of problem under consideration. Subramanian, Pekny, and Reklaitis (2000) and Pekny (2002) provide a detailed discussion.

APCI's EO was developed in collaboration with APC using APC's *VirtECS Design* tool. The model parameter input database and basic structure of the NO_x problem were developed at APCI. APC then created a customized *NOx Model Manager* and tailored the problem formulation and solution algorithm to perform the optimization in an acceptable amount of time using their proprietary *VirtECS* solver technology.

The *NOx Model Manager* determines the lowest cost of compliance with the Texas State Implementation Plan. First, the EO performs an enterprise-wide optimization, where all companies act independently. This is followed by a region-wide optimization, where all participating companies act cooperatively. The optimization determines at each time period which companies should over-control their point sources to what extent to generate and sell credits, and which companies should under-control their point sources to what extent and buy credits.

Nonlinear Programming Component

Trading is implicit in the MILP framework. The EO performs a subsequent NLP optimization based on the MILP results to determine the tons per year (TPY) and dollar trades corresponding to the Cap & Trade program. APCI developed this *NOx Trading Model* using the *Generalized Algebraic Modeling System* (*GAMS*), described by Brooke *et al.* (1998).

The difference in capital investment between the two MILP solutions forms the economic basis for determining the marginal values of the buyers and sellers ($/TPY). The difference in the marginal values provides the driving force for trades. The net savings between the region- and enterprise-wide MILP solutions is partitioned between the sellers and buyers through the trades. The NLP trading model essentially maximizes the area between the buyer and seller trading curves, which corresponds to the net capital savings. The sellers receive more revenue than their additional investment required to generate credits, and buyers pay less through credit purchases than their alternative to invest in additional control technology.

The difference in margins for each trade represents the "negotiating range" for the individual trades, so there are many potential solutions to this problem. Many of these would be impractical from a commercial standpoint, because their interpretation is for a seller to trade at the break-even cost for the additional investment or a buyer to trade at the break-even point for the alternative investment. Therefore additional constraints are added to drive the solution towards a sharing of the savings within the individual trades. A high-level abstraction of the NLP problem is:

Minimize *Disparity in Shared Savings*

Subject to:
Total TPY Bought = Sum TPY Traded by Sellers
Total TPY Sold = Sum TPY Traded by Buyers
Dollars Bought = Sum Dollars Traded by Sellers
Dollars Sold = Sum Dollars Traded by Buyers
Dollars Sold ≥ Seller's Additional Investment
Dollars Bought ≤ Buyer's Alternative Investment
Buyers' Trade Margin ≥ Seller's Trade Margin
Total Savings Distributed Through Trades
Definition of Shared Savings

Results

These models have been solved for the entirety of the point sources belonging to all the companies in the HGA. Activity level and emission factors were taken at the base-values from the TCEQ database. Marketing reports and internal studies were used to generate appropriate treatment technology options, emission-reduction capabilities, and costs for each point source.

Figure 1 illustrates the 15.7% cost savings between a region-wide optimization with Cap & Trade compared to an enterprise-wide optimization. Figure 2 illustrates an example of a trading curve between buyers and sellers.

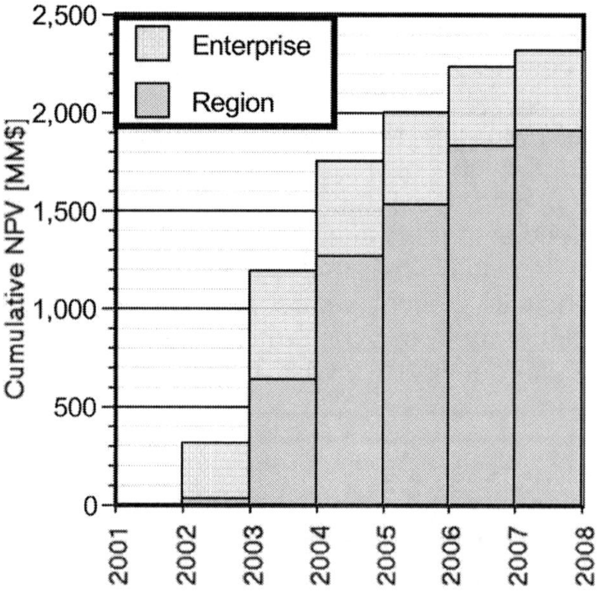

Figure 1. *Cost Savings*

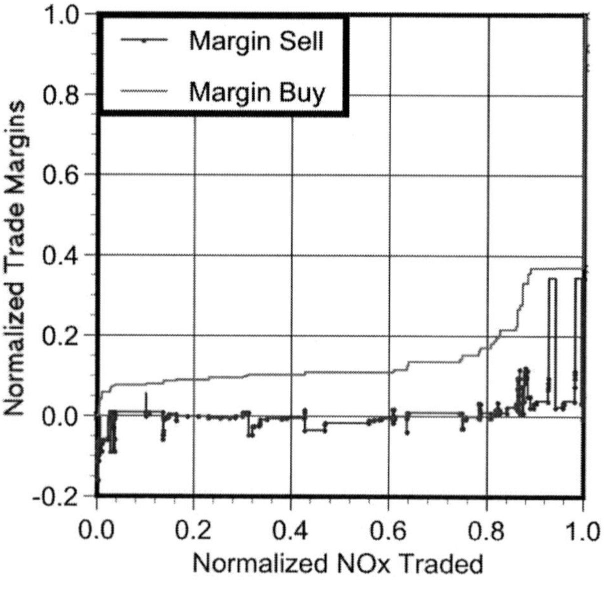

Figure 2. *Normalized Trading Curve*

The ECCA solution returns the enterprise-wide EO solution directly to the customer through the web-based supporting software. APCI's sales team follows up with a benefits opportunity analysis to those companies that can achieve a significant cost savings through trading.

Customer Input Data and Interface

The customer inputs a monthly forecast of activity level and fuel source for each point source. The allowances are based on the data from the TCEQ database. The customer then provides one or more (or no) control technology options under consideration as well as any control technology selections that have already been decided, along with the timing of those decisions. Each control technology requires a cost and emission-reduction capability. Two technology options may be combined, where the performance of one is dependent on the selection of the other. For example, the performance of a Selective Catalytic Reduction (SCR) unit will be different if it is connected directly to the point source or if it is downstream of a Low NO_x Burner (LNB).

Unique features have been added to the EO model to match specific company requirements and regional regulations. Companies can identify equipment that must be reduced (even if non-optimal) due to other reasons such as consent decrees or plant efficiency projects. They can specify whether or not to roll over extra allowances at the end of each year. Depending on the local regulations, companies can use Emission Reduction Credits and Discrete Emission Reduction Credits (ERCs and DERCs) to add to their allowance limits. These additional allowances can be continuously added to the solution. Finally, they can restrict the installation of technology until a plant outage occurs. This is done on a unit-by-unit basis to allow the greatest degree of flexibility. Figure 3 illustrates one of the customer screens.

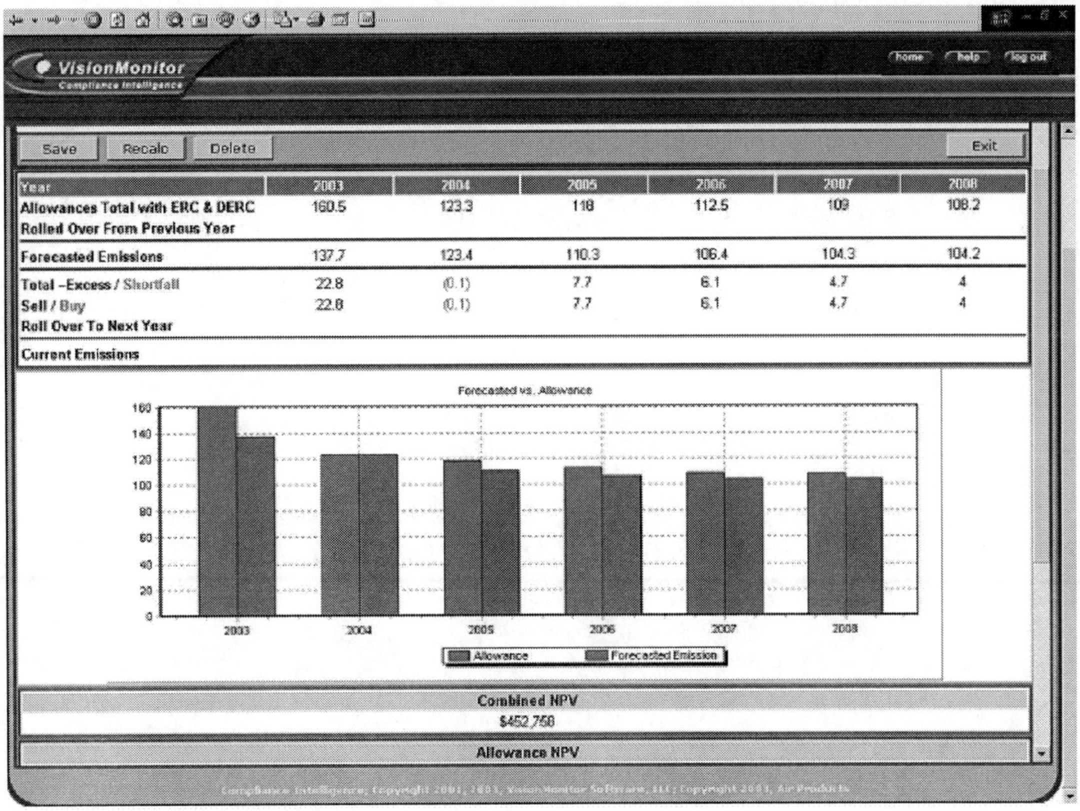

Figure 3. *Example of a Customer Display Screen*

Conclusions

APCI's ECCA solution is a unique environmental compliance monitoring and optimization product designed to deliver value-added services to our customers. The EO is a key differentiating component of this Services & Solutions Business offering. The EO couples APCI's in-house modeling and optimization expertise with APC's proprietary *VirtECS* solver technology.

The EO provides a successful example of how technical advances made by academic researchers can be transformed into viable commercial products. By cultivating a long-term industrial-academic relationship, changing business needs can be used to identify new opportunities to advantageously apply optimization technology to determine solutions to critical problems.

References

Brooke, A., D. Kendrick, A. Meeraus, R. Raman, and R.E. Rosenthal (1998). *GAMS: A User's Guide.* GAMS Development Corporation, Washington, DC.

D'Aquino, R.. (2003). The NO_x Market: Play It Like a Pro. *Chemical Engineering Progress*, **99**(6), 9-12.

Huston, R.J., R.B. Marquez, K.H. White, and J.A. Saitas (2002). *Mass Emission Cap & Trade Program (MECT)*, Texas Natural Resource Conservation Commission, Austin, TX.

Pekny, J.F. (2002). Algorithm architectures to support large-scale procss systems engineering applications involving combinatorics, uncertainty, and risk management. *Computers and Chemical Engineering*, **26**, 239-267.

Subramanian, D., J.F. Pekny, and G.V. Reklaitis (2000). A simulation-optimization framework for addressing combinatorial and stochastic aspects of an R&D pipeline management problem. *Computers and Chemical Engineering*, **24**, 1005-1011.

Paper 21A

Operational Excellence: Best Practices in Manufacturing

Steve Williams
Vice President, Product Management
Manufacturing Execution Systems/Integration

Aspen Technology, Inc.

Prepared for presentation at

AIChE Spring Meeting

April 2004

Session T3

April 2004

Unpublished

Operational Excellence:
Best Practices in Manufacturing

by
Steve Williams
Vice President, Product Management
Manufacturing Execution Systems/Integration
Aspen Technology, Inc.
Houston, Texas

Abstract

This paper considers the role of new and emerging technologies in helping companies to achieve operational excellence. Among those key technologies are the following:

- Integration infrastructure for bringing together diverse data in manufacturing environments
- Improved visibility (web-based views of real-time information)
- Event management for intelligent alerting and automated responses to important business events
- Performance scorecarding and analytics (for making performance-vs-target results visible and providing the analysis environment to understand and improve performance)
- Look-ahead analytics for projecting future performance and improving future profitability (not only from learning but also from past performance).

These technologies are available today and can be implemented by process industry companies as they strive to establish and maintain operational excellence in manufacturing.

Introduction

For more than twenty years, manufacturers have focused on improving the quality of their processes as a way to achieve operational excellence. Advancements in technology over the past two decades have played a key role in driving the growing number of quality improvement initiatives in manufacturing worldwide, and have underpinned the more recent Six Sigma projects undertaken in the United States.

Today, however, process manufacturers are recognizing that operational excellence relies not only on traditional technology, but also on the effective combination of emerging technologies and the role of the "human factor." Over-emphasis of one aspect to the detriment of the other will limit an organization from achieving its true potential.

A suite of software applications has been developed in order to fulfill both the technical and human requirements of achieving operational excellence. This emerging technology is available today and can be implemented to provide process industry companies with capabilities that improve visibility, manage events, and provide performance scorecarding and analytics for improving and understanding performance.

Background and Historical Context

In discussing the current technologies designed to help process manufacturers achieve operational excellence, it is important to understand the breakthroughs and the legacies that have created today's operating environment.

Technology and PC-Based Tools

The first application of technology to specific areas within the process industries occurred in the 1970s, when Distributed Control Systems (DCS) provided operators with improved visibility of a particular process and an unprecedented easy-to-use environment. In the continuous process industries, advanced control technology, such as Dynamic Matrix Control (DMC), successfully enabled process manufacturers to achieve improvements in productivity and yield by pushing processes closer to operating constraints. Later, the advent of process historians offered operators and selected operations staff a standard tool for investigating performance issues and understanding process performance.

Augmenting these technologies over the years has been the emergence of PC-based tools for a variety of functions, including data analysis. Microsoft Excel, for example, is widely deployed in many plants, although all too often it is deployed in "information silos" that offer very little sharing of information and lead to localized decision making. Further technologies that address plant scheduling, planning, and execution have also emerged. With all these systems have come masses of data…so much so, in fact, that manufacturers are heard to say that they are "drowning in data." What they really mean, however, is that they need to distill *information* from the abundance of *data*—and that often involves correlating data from many different systems.

As underlying systems have increasingly become commodities, and as operating environments have become populated with disparate systems, it is clear that integration of the different technologies is an important issue if key business processes are to be automated. This requirement has become even more acute as manufacturers strive for more agile decision-making, based on an accurate understanding of the current state of process operations.

The Impact of ERP Systems

In the early 1990s, the advent of enterprise resource planning (ERP) systems provided a comprehensive, integrated source of information for the entire organization, thereby allowing changes to propagate easily across a business. ERP systems have indeed achieved benefits in operational excellence, but they have also incurred large investment costs—costs so large, in fact, that companies are today reluctant to repeat the experience. The ERP systems have also established some important standards which customers expect suppliers in the manufacturing space to emulate.

Integration Technologies and Standards

To solve to the complexities of the integration of different business systems, information technology (IT) departments have sometimes selected enterprise application integration (EAI) technologies. EAI suppliers such as IBM, webMethods and TIBCO have developed powerful "integration stacks" specifically designed to make systems integration easier and more reliable. While these general integration technologies do not necessarily take advantage of some of the particular characteristics of the process industries, they have brought some potential order to the technology environment for process manufacturers. Order is very much required—but lack of order and lack of standards characterize the current environment.

The Human Factor

As technologies evolved over the past two decades, they did so against a backdrop of mandated cost and manpower reductions. Exacerbating this problem is the fact that many experienced operations personnel are about to retire. This situation, combined with staffs that have been consistently reduced for many years, creates a critical knowledge management challenge.

Operational Excellence in Manufacturing

Operational excellence may be achieved by a series of steps, as discussed below.

Step 1: Visibility and Empowerment

Process manufacturers must provide users across the organization with meaningful, real-time visibility across systems and operational silos to better contribute to organizational goals. In many cases, this requirement has meant leveraging new applications and technologies, such as web-based technology. Web-based technology brings new possibilities to operations; for example,

- It enables all Operations personnel within a plant to see one "version of the truth," thereby helping to overcome an organization built on silos. It encourages collaborative decision making and optimization of operations on a broader basis.
- It can also support this "version of the truth" vertically, and not simply horizontally, within an organization. Supervisors, managers, and operators can all be provided with appropriate operational views to help them to make better decisions.

ARC Advisory Group has described this new generation of web-based visibility tools as "Performance Personalization Tools" (PPTs) because they employ web-based technology—underpinned by a data model—allowing data to be stored, retrieved and navigated in context. The net effect is to convert masses of data into information, which can be intuitively accessed, to greatly facilitate decision-making.

Step 2: Business Response and Agility

Process industry manufacturers must make high quality and rapid decisions. **Figure 1** depicts the "latency period," defined as the delta between the impact of time to react to a business event on the value of the action taken. Agility of response is important at all levels of operations, from the plant to the business. Intelligent alarm management and escalation are key to reducing the latency period and thereby increasing the value of rapid, accurate decisions to the business. Operators will often acknowledge alerts as a matter of course; however, situations can occur in which a particular combination of alerts from the plant, or a composite alert (combining alerts from the DCS with those from a reliability system, for example) signifies an impending larger problem which may have a major impact on plant reliability. It is imperative to be able to flexibly configure systems to handle such hybrid scenarios.

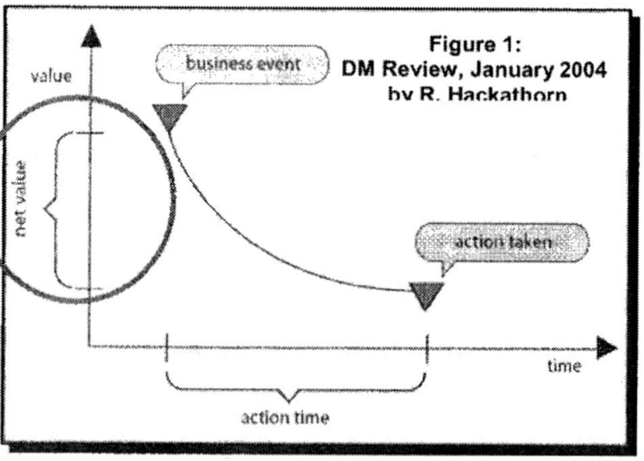

Figure 1: DM Review, January 2004 by R. Hackathorn

Escalation capabilities are also important. If a particular alert is not getting the attention it deserves after an appropriate time (or number of occurrences), for example, it should be escalated to a higher level in the supervisory hierarchy so that it is addressed. From a business perspective, agility at a higher level allows companies to manage important events in the supply chain, such as significantly increased demand from a particular customer. When these business events can be coupled with a specific scenario analysis, companies can anticipate the best course of action by projecting likely future performance.

Key aspects of response agility in the process industries include collecting process knowledge, embedding process knowledge, automating responses in particular scenarios, and enabling all appropriate event sources.

Comprehensive operational excellence systems will also have the capability to automate certain aspects of business response, for example, by passing information from one system to another automatically or triggering a sequence of response actions.

Step 3: Integration and Standards

A typical process operations environment consists of many systems, each of which contains important data. Often, however, these rarely-integrated systems are connected with a variety of point-to-point interfaces, or may even be comprised on standalone spreadsheets for a small set of users.

The biggest challenge for point-to-point integration is the need to adapt to change. A change in an application can affect a number of interfaces: the ripple effect can be large and expensive. A new, more flexible solution that copes as the business and information requirements change is fundamental in enabling operational excellence. Any new approach must also cope with the practical realities of integration in modern environments: namely environments which are distributed and where networks and applications occasionally fail.

The roll out of SAP across many companies places requirements on data integration with the plant floor. Information about production orders is required in the plants and information about actual production is required in SAP. These information requirements increase in complexity when other systems, such as supply chain management, are introduced. Given these growing requirements, operational excellence can be achieved only through the combination of bi-directional business process execution as well as visualization.

Step 4: Business and Operational Alignment with Scorecards and Dashboards

Most Operations personnel working in a large plant have little idea of the official corporate strategy; nor do they know how they can best maximize their contribution to that strategy. Similarly, visibility of operational performance between the corporate and business levels is often very poor. A scorecarding system (perhaps based on a balanced scorecard approach) can be very effective in making performance visible, thereby forcing an emphasis on visibility of performance targets and actual performance and encouraging an environment of increased accountability.

Many companies have undertaken the non-trivial task of defining KPIs which are important to them (some have created executive dashboards that typically provide visualization of business information). Those that have are in a good position to adopt scorecarding technology, particularly if these have been defined across multiple plants within a business unit. For those that have not taken the adoption challenge, the task is more significant. The important interplay of human and technology factors is very evident in this situation. To overcome these challenges, companies should work with a vendor who has both the domain expertise as well as the technology to apply to this challenge.

At any rate, the scorecarding system should be web-based to ensure maximum performance visibility and use. It should support hierarchically-defined key performance indicators (KPIs) so that users can "drill-down" into areas of poor performance to understand the contributing factor(s) causing performance targets to be missed. An advanced scorecarding system supports not just targets and actuals, but also potential performance targets. These are maximum potential performance levels, which might be computed via on-line models for example.

Achieving Operational Excellence in Manufacturing

A suite of software applications, Aspen Operations Manager, has been developed in order to fulfill both the technical and human requirements of achieving operational excellence. This suite is available today and can be implemented to provide process industry companies with capabilities that fulfill the demands of the steps outlined above.

Visibility and Empowerment: Role Based Visualization

The Role Based Visualization application incorporates a variety of tools to address the following tasks:

- Building web-based graphics. Some of these authoring tools are themselves web-based, and therefore enable user-configurable environments. The tools include libraries of standard "drag and drop" process graphics elements to create web pages.
- Converting DCS graphics (e.g., Honeywell TDC 3000) to a web implementation, encapsulating the graphic in the web
- Organizing data storage, retrieval and navigation, thereby effectively converting the data into information
- Offering a portal environment that allows different graphic elements to be composed and screens to be tailored to the varying needs of operational personnel.

Figures 2 and **3** represent Role Based Visualization screens.

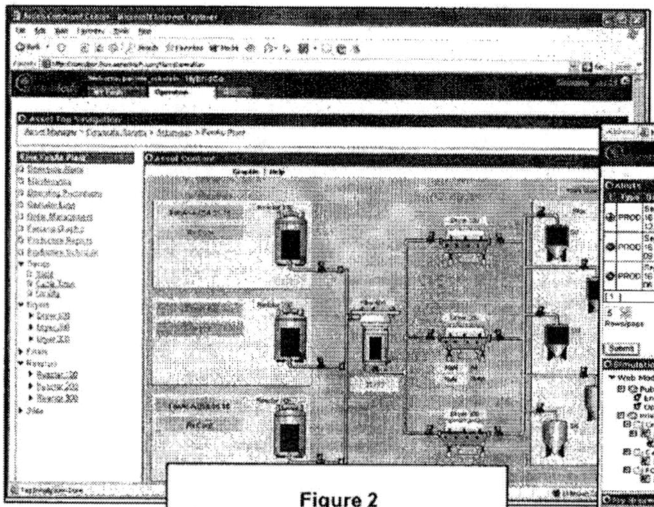

Figure 2
Context-sensitive content

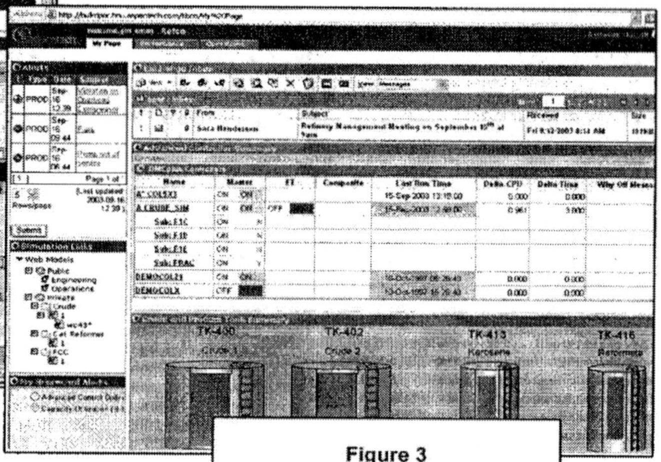

Figure 3
Hybrid systems environment

Business Response and Agility: Event Management

To comprehensively address the needs of the process industries, the Event Management application incorporates a number of sub-systems, including the following:

- Portal-packs to deliver web content into a portal environment.
- A scalable Rules Engine, configurable with wizards for most users or with a full development environment for power users.

- Ability to template specific solutions.
- Integration with a broad variety of standard components and fully compliant with Aspen Operations Manager - Integration Infrastructure.

The Event Management application is often used in conjunction with other business and plant systems. **Figure 4**, for example, depicts how Event Management can complement a compliance management system. The application provides suggestions for the likely cause for specific limit violations, as well as advice on what steps should be taken to correct the violation. Event Management may also be used to trigger workflows involving manual and automated steps.

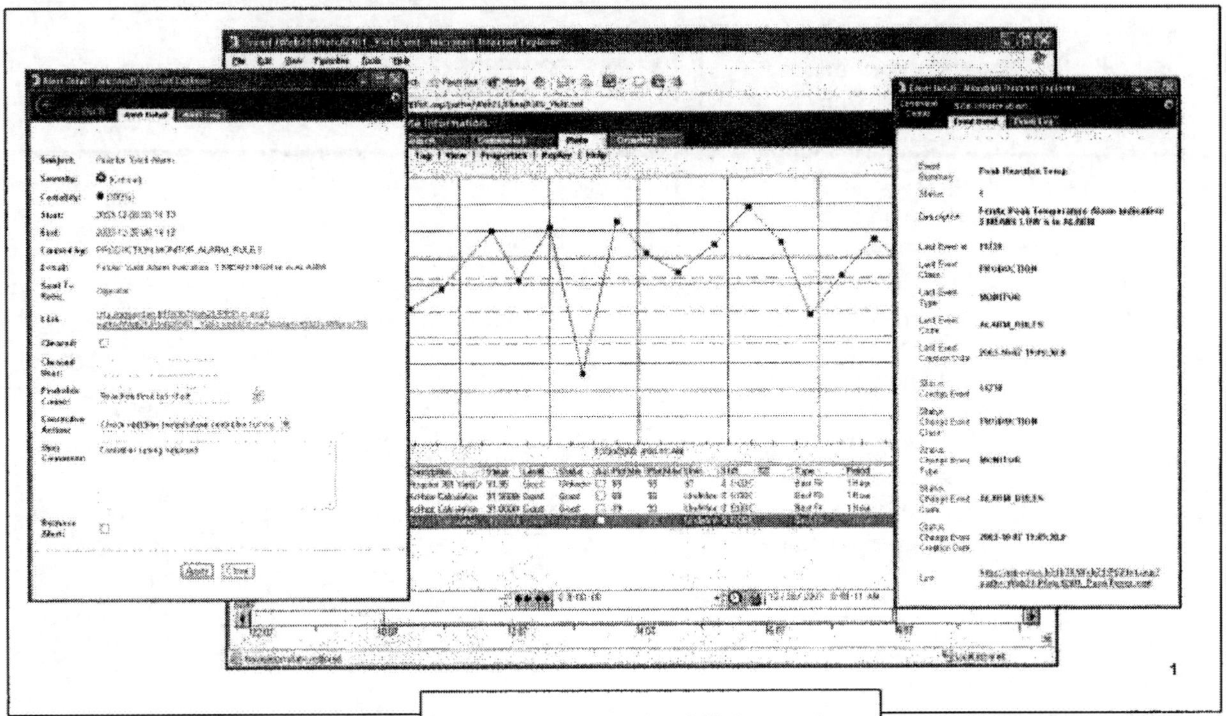

Figure 4: Event alert drill-down and embedded knowledge

Integration and Standards: Integration Infrastructure

The Integration Infrastructure embeds a robust EAI architecture from TIBCO and augments it with a data model and content and connectivity for the process industry, aligned with the ISA S95 standard. Standard messages for plans, schedules, production balances, and process values have been defined, as well as components for the Master Data Manager module. These components—which include units of measure, material codes, properties, prices, equipments, and hierarchy—allow a central point of definition and synchronization with other Master Data systems, and also provide a means of propagating configuration information to a number of applications.

Business and Operational Alignment: Performance Scorecarding

The Performance Scorecarding application is a powerful application that stores data in a structured, dimensioned form which enables rapid "slice and dice" analysis and enables conditions to be defined for variables. These conditions can then be linked to traffic lights to graphically flag significant deviations. An example might be in a polymer process where a transition time alert is generated within a business. Using the Performance Scorecarding applications, users can drill-down to establish that the problem is occurring at a particular production facility for a particular transition. Other key features of the Performance Scorecarding application include the following:

- Powerful reporting (including HTML reports)
- Ability to link to a variety of relational data-sources
- Ability to support drill down
- Ability to support 'drill-through' (connections to sources of data for display of more granular data)
- Compatibility with Microsoft Excel and AspenTech's Integration Infrastructure
- Pre-configured data structures for the process industry

Figure 5 shows an example of a typical Performance Scorecarding environment and the ability to drill down in a particular area to display key results graphically.

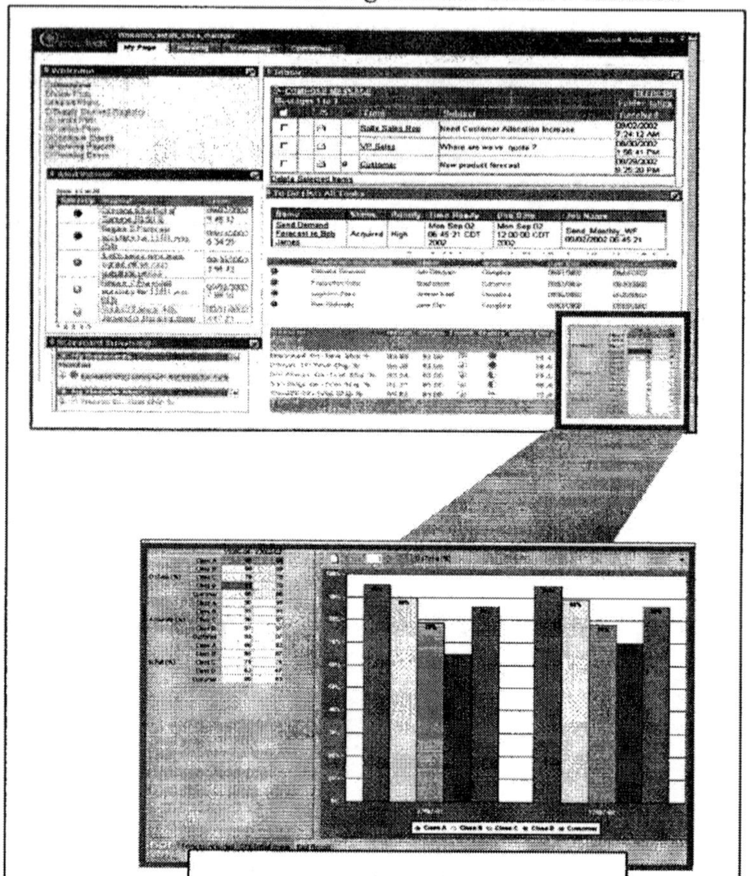

Figure 5
Typical scorecarding environment

Conclusion

For process manufacturers today, the new technology that will help achieve operational excellence is entering the market now. With increasing pressure from shareholders, customers and even employees to ensure the business performs to the stated goals, these technologies can help achieve a level of collaboration and excellence not seen in the past. In many cases, the infrastructure is in place. Even with cost pressures these technologies can help organizations achieve the operational excellence they need, with their existing resources (infrastructure and talent) as long as they act now.

Solutions for Successful Chemical Inventory Management
By Marc E. Dillon, P.G., Chemical Management Products Manager, ESS

Introduction

Chemical inventory management can be a daunting task for companies of all sizes. The responsibility typically falls squarely into the hands of Environmental, Health & Safety (EH&S) or material distribution staff that may be operating with limited resources. These professionals are constantly concerned with maintaining and demonstrating compliance with their facility permits and other applicable corporate, federal, state and local policies and regulations.

One element of demonstrating compliance requires the preparation of various reports that require knowledge of the type, location, storage and usage of chemicals in facilities. In order to be able to accurately generate these reports, chemical process engineers or EH&S professionals must have access to accurate chemical inventory management data. The old adage, "you can't manage what you don't measure" has never been more true than with chemical inventory and other elements of environmental management.

This article reviews of the types of software products that are available to assist EH&S professionals manage, automate, and improve their overall regulatory compliance in regard to chemical inventory processes. Recommendations are provided about when certain types of software products might be used and several key issues and questions are provided to assist chemical process engineers and EH&S professionals evaluate potential chemical inventory management software vendors. In addition, some "lessons learned" from system implementations are offered so that these pitfalls may be avoided once the system is selected and purchased.

Chemical inventory management essentially consists of the following components:

1. Chemical purchases and receipts
2. Chemical inventory transactions (storage and use)
3. Inventory Management
4. Record keeping
5. Reporting

These five elements comprise the chemical inventory management life cycle: from when chemicals are ordered, purchased and received at the facility, to when they are stored, used and depleted for disposal. The first two elements deal with the identity and quantity of materials entering the facility or enterprise. Inventory management includes tracking and maintaining proper inventory levels to maintain production or other important operational processes. Record keeping includes maintaining accurate records of the previous elements and up-to-date material composition, properties, regulatory information, and Material Safety Data Sheets (MSDSs) for the chemicals in inventory. All these data are used for creating reports including SARA 312 Tier I or Tier II, and permit required reports that may include, among others, hazardous air pollutant (HAP) and volatile organic compound (VOC) usage data.

Finding the Right Fit: Features of Chemical Inventory Management Systems

Those responsible for chemical inventory management for their facility or corporation typically employ various software products to meet the challenge. The complexity of chemical management software solutions must reflect an organization's needs, which are determined by its goals, size, operational processes and the number and type of regulatory requirements to which the organization must comply.

Chemical inventory management software products typically fall into three basic categories

1. Chemical distribution inventory tracking.
2. Transaction-based chemical inventory tracking with integrated access to regulatory requirements and report generation capabilities.
3. Enterprise level, transaction-based chemical inventory tracking with integration to other environmental monitoring systems such as waste, water, or air.

The first category of products is concerned only with data related to the occurrence and distribution (location) of chemicals and is generally stored in a spreadsheet or database. The data is typically updated a few times a year as chemicals and their locations are added or removed. This static inventory does not track the day-to-day inventory transactions. It simply tracks the type and maximum and average quantity of a chemical at each location of a facility.

The advantages of this type of software are low cost, ease of use and installation. The disadvantages include a lack of real-time inventory control, limited report generation capability, no actual chemical usage data — only estimates — and limited or no integration capability with external data sources, such other environmental monitoring systems or with other information technology (IT) infrastructure.

Transaction-based chemical inventory tracking provides tracking of batches or individual chemical containers. The transactions track the occurrence, distribution and movement of each chemical container or batch of containers. With transaction systems, data such as type and quantity of chemical used, location of use, how the chemical is applied (spray, brush, etc.) and person or cost center using the chemical can be captured.

Furthermore, with a serialized, bar-coded inventory system, data elements such as expiration date, batch, and lot numbers can be tracked to each individual container.

The advantages of transaction-based systems include providing complete, real-time inventory control; actual or estimated chemical usage data; container labeling capabilities; container expiration date tracking; standard and customized reporting formats — often with electronic submittal to agencies, if appropriate — and links to external data sources. Another advantage of these systems is the ability to allow remote data to be collected with handheld devices and imported into the inventory database. Most of these systems provide a link between the chemicals and their associated regulatory requirements from federal, state and local agencies, and can link to the chemical's MSDS in text and image format. These solutions are available in both browser-based and client-server based formats.

The disadvantages of this type of software are its higher purchase cost and potential challenges with implementation. Since these systems often involve IT departments, management and other stakeholders, such as material distribution and procurement departments, the groups' competing priorities, conflicting workflow and data flow processes can make the project challenging. Implementation time (and costs) can be very high depending upon the quality of existing data prior to startup and the amount of configuration needed to match workflow.

Enterprise software includes all the features of transaction-based systems, but includes integration with other key environmental management functions at an enterprise level within an organization. This means that other EH&S systems, such as waste management, air emissions tracking, compliance management, storm water, wastewater, industrial hygiene, and occupational health & safety can be integrated with chemical tracking.

The advantages of an enterprise chemical inventory tracking system is that because a facility's emissions to air, waste and water originate from the chemicals it maintains in inventory and the processes that use those chemicals, enterprise software provides a true mass balance, from cradle to grave, of all materials touched by a single facility or across an entire corporation. These solutions are typically deployed in browser-based formats.

The disadvantages of enterprise software are its re relatively high cost and challenges in implementation. Often, implementation of these systems requires a company to clearly define and refine their workflow and dataflow processes and then migrate existing data into the application database and model these processes during configuration. However, this usually becomes an advantage for the system's use as efficiencies are maximized. Implementation also involves numerous stakeholders, including IT management and infrastructure, material distribution, procurement, accounting, and site and corporate level management.

The benefits of implementing and operating an efficient chemical inventory management system most often far outweigh the purchase, implementation and support cost. These benefits include:

- Accurate data for EH&S, material distribution and management departments
- More efficient use of material and time
- Consolidation of chemical inventories and lower chemical purchase costs
- Reduced chemical usage
- Waste minimization
- Pollution prevention
- Ability to perform "first in, first out" (FIFO) or "just in time" (JIT) inventory control
- Internal and external customer satisfaction
- Concentration on core business
- Enhanced relationship with regulatory agencies

Evaluating Chemical Inventory Management Suppliers

Many vendors offer the three types of products just described. However, before choosing a vendor, an organization must do some homework. Those professionals in charge of selecting, purchasing, and using the system must ask:

- What are the specific goals and objectives for using such a system?
- What are the "must have" and "nice to have" features and functionality of the selected system?
- What is the specific, expected output of the system?
- Is a stand-alone chemical inventory system or multi-module (air, water, waste, etc.) system needed?
- Is integration needed with external systems (ERP or other)? If so, what other systems and why?
- Will sufficient internal expertise and resources be available to implement all or part of the system or will one rely on the software vendor or a third-party implementation partner(s)?

Questions to ask potential suppliers should include:

- Does the firm have an understanding of your industry and your specific facility?
- How many systems have they implemented in the past year or two?
- Do they have a quality assurance, quality control policy and plan?
- What is the depth and availability of technical support?
- How many experienced project managers, developers and programmers are on staff? What are their qualifications and backgrounds?
- What is the company's financial history?
- Is the vendor willing to provide references of companies who recently purchased or implemented their software, and give potential clients the right to solicit feedback from them?
- What is the vendor's experience with the applicable regulations and management practices concerning your operation?
- What types of databases does their software support?

Lessons Learned

The following are some key points to consider before selecting and purchasing a chemical inventory tracking system. These are not trivial points, as history can point to many implementations that have faltered and expectations not being met because these lessons learned were not followed.

- Ensure that all stakeholders are involved in the process of gathering and developing system requirements. Otherwise, actual and expected system functionality will not match and implementation difficulty (and cost) will escalate.
- Ensure that there is both an IT and business (EHS) stakeholder involved at the minimum.
- Collect system hardware and software requirements from the potential vendors before making a decision to ensure that there are no hidden costs and that compatibility will not be an issue.
- Collect references from the vendor for other customers, get permission to contact them and follow up (without the vendor) for feedback.
- Arrange for system demonstrations with the vendor and ensure that product managers and technical staff, in addition to sales staff, are present for questions.
- If data is to be migrated into the new system, review the data to ensure completeness and quality. Be able to give specifics on the data source (database type and version, etc.) If integration is needed with outside systems (like ERP system), coordinate with those stakeholders to provide system specifics to the chosen vendor and/or implementation partner.

- If processes, workflow, or dataflow are scheduled for change or need to be changed, ensure that these are well documented and reviewed prior to software selection and implementation planning.
- Develop a plan to calculate and evaluate the business value and return on investment (ROI) expected. This may include baseline values for chemical purchases, etc. at a given production level before and after system implementation and use.

Conclusion

Chemical inventory management software can provide environmental compliance data to assist professionals with a multitude of regulatory requirements that need to be met within any organization. Selecting and implementing the appropriate type and size of solution is critical and dependent on the goals and size of the operation and the quantity of its requirements. Properly selecting the right system can provide value and save a company time and money far in excess of a solution's purchase and implementation costs.

Mr. Dillon is the chemical management products manager and director of consulting services for ESS, headquartered in Tempe, Ariz. Mr. Dillon is a professional geologist and has more than 15 years of experience in environmental management and engineering, in the aerospace industry, and in consulting.

American Institute of Chemical Engineers
3 Park Avenue
New York, New York 10016-5991

ISBN: 978-1-60423-474-9

American Institute of Chemical Engineers

Conference on Emerging Energy Frontiers in Research and Innovation

Topical Conference at the
2008 AIChE Spring National Meeting

April 6-10, 2008
New Orleans, Louisiana, USA